Lecture Notes in Computer Science   7827

Commenced Publication in 1973
Founding and Former Series Editors:
Gerhard Goos, Juris Hartmanis, and Jan van L ͡e͡u͡w͡e͡n

Bonghee Hong   Xiaofeng Meng   Lei Chen
Werner Winiwarter   Wei Song (Eds.)

Database Systems
for Advanced Applications

18th International Conference, DASFAA 2013
International Workshops: BDMA, SNSM, SeCoP
Wuhan, China, April 22-25, 2013
Proceedings

 Springer

Volume Editors

Bonghee Hong
Pusan National University, Pusan 609-735, South Korea
E-mail: bhhong@pusan.ck.kr

Xiaofeng Meng
Renmin University, Beijing 100872, China
E-mail: xfmeng@ruc.edu.cn

Lei Chen
Hong Kong University of Science and Technology
Kowloon, Hong Kong
E-mail: leichen@cse.ust.hk

Werner Winiwarter
University of Vienna, 1090 Vienna, Austria
E-mail: werner.winiwater@univie.ac.at

Wei Song
Wuhan University, Wuhan 430072, Hubei Province, China
E-mail: songwei@whu.edu.cn

ISSN 0302-9743   e-ISSN 1611-3349
ISBN 978-3-642-40269-2   e-ISBN 978-3-642-40270-8
DOI 10.1007/978-3-642-40270-8
Springer Heidelberg New York Dordrecht London

Library of Congress Control Number: 2013934238

CR Subject Classification (1998): H.3, H.4, I.2, C.2, H.2, H.5, D.2

LNCS Sublibrary: SL 3 – Information Systems and Application
incl. Internet/Web and HCI

Typesetting: Camera-ready by author, data conversion by Scientific Publishing Services, Chennai, India

Printed on acid-free paper

Springer is part of Springer Science+Business Media (www.springer.com)

Preface

Database Systems for Advanced Applications (DASFAA) is an annual international database conference, located in the Asia-Pacific region, which showcases state-of-the-art R&D activities in database systems and their applications. It provides a forum for technical presentations and discussions among database researchers, developers, and users from academia, business, and industry. DASFAA2013, the 18th in the series, was held in Wuhan during April 22–25, 2013. Among the proposals submitted in response to the call-for-workshops, we carefully selected three workshops, each focusing on a specific area that contributes to the main themes of the DASFAA conference. This volume contains the papers accepted for these three workshops that were held in conjunction with DASFAA 2013:

- The First International DASFAA Workshop on Big Data Management and Analytics (BDMA 2013)
- The 4th International Workshop on Social Networks and Social Web Mining (SNSM 2013)
- The International Workshop on Semantic Computing and Personalization (SeCoP 2013)

We are very grateful to the workshop organizers for their tremendous effort in soliciting papers, selecting papers by peer review, and preparing attractive programs. We asked all workshops to follow a rigid paper-selection process, including the procedure to ensure that any Program Committee members were excluded from the review process of any paper they are involved in. A requirement about the overall paper acceptance rate was also imposed on all the workshops. We would like to express our appreciation to Jianzhong Li, Zhiyong Peng, Qing Li, and many other people for their support in the workshop organization. Our thanks also go to Cyrus Shahabi, Ge Yu, Guandong Xu, Lin Li, and Yi Cai for their hard work in organizing these workshops.

April 2013

Hong Bonghee
Xiaofeng Meng
Lei Chen
Werner Winiwarter
Wei Song

DASFAA 2013 Workshop Organizer

Workshop Co-chairs

Hong Bonghee	Pusan National University, South Korea
Xiaofeng Meng	Renmin University, China
Lei Chen	Hong Kong University of Science and Technology, China

Publication Co-chairs

Werner Winiwarter	University of Vienna, Austria
Wei Song	Wuhan University, China

First International DASFAA Workshop on Big Data Management and Analytics (BDMA 2013)

Workshop Co-organizers

Cyrus Shahabi	University of Southern California, USA
Ge Yu	Northeastern University, China
Hong Bonghee	Pusan National University, South Korea

Program Committee

Peter Baumann	Jacobs University Bremen, Germany
Taek Chung	Humanvest.co, USA
Xiaoyong Du	Renmin University, China
Hong Gao	Harbin Institute of Technology, China
Wook-Shin Han	KyungPook National University, South Korea
Jinho Kim	Kangwon National University, South Korea
Seon Ho Kim	University of Southern California, USA
Joonho Kwon	Pusan National University, South Korea
SangKeun Lee	Korea University, South Korea
Sang-goo Lee	Seoul National University, South Korea
Yoon Joon Lee	KAIST, South Korea
Zhanhuai Li	Northwestern Polytechnical University, China
Hyoseop Shin	Konkuk University, South Korea
Ha-Joo Song	Pukyong National University, South Korea
Guoren Wang	Northeastern University, China
Jongwook Woo	California State University, USA
Jae Soo Yoo	Chungbuk National University, South Korea
Aoying Zhou	Eastern China Normal University, China

4th International Workshop on Social Networks and Social Web Mining

Workshop Co-chairs

Guandong Xu	University of Technology, Sydney, Australia
Lin Li	Wuhan University of Technology, China

Program Committee

Nitin Agarwal	University of Arkansas, USA
Toshiyuki Amagasa	University of Tsukuba, Japan
James Bailey	University of Melbourne, Australia
Kevin Chai	IBM, China Research Lab, China
Ling Chen	University of Technology, Sydney, Australia
Peter Dolog	Aalborg University, Denmark
Flavius Frasincar	Erasmus University Rotterdam, The Netherlands
Irene Garrigos	University of Alicante, Spain
James Geller	New Jersey Institute of Technology, USA
Yanhui Gu	University of Tokyo, Japan
Hyoil Han	Drexel University, USA
Hiroyuki Kitagawa	University of Tsukuba, Japan
Carson Kai-Sang Leung	University of Manitoba, Canada
Wenxin Liang	Dalian University of Technology, China
Mukesh Mohania	IBM India Research Laboratory, India
Satoshi Nakamura	Kyoto University, Japan
Tieyun Qian	Wuhan University, China
Sherif Sakr	NICTA, Australia
Munehiko Sasajima	Osaka University, Japan
Kazutoshi Sumiya	University of Hyogo, Japan
Xiaohui Tao	University of Southern Queensland, Australia
Chaokun Wang	Tsinghua University, China
Daling Wang	Northeastern University, China
Jianmin Wang	Tsinghua University, China
Zongda Wu	Wenzhou University, China
Zhenglu Yang	University of Tokyo, Japan
Junjie Yao	Peking University, China
Hwanjo Yu	POSTECH, South Korea
Jianwei Zhang	Kyoto Sangyo University, Japan
Xiuzhen Zhang	RMIT, Australia
Bin Zhao	East China Normal University, China
Yu Zong	West Anhui University, China
Lei Zou	Peking University, China

International Workshop on Semantic Computing and Personalization

Workshop Organizer

Yi Cai   South China University of Technology, China

Program Committee

Jian Chen   South China University of Technology, China
Li Chen   Hong Kong Baptist University, China
Yunjun Gao   Zhejiang University, China
Raymong Y.K. Lau   City University of Hong Kong, China
Rong Pan   Sun Yat-Sen University, China
Shaojie Qiao   Southwest Jiaotong University, China
Derong Shen   Northeastern University, China
Shaoxu Song   Tsinghua University, China
Yuqing Sun   Shangdong University, China
Jie Tang   Tsinghua University, China
Raymong Wong   Hong Kong University of Science and
                                     Technology, China
Haoran Xie   City University of Hong Kong, China
Jing Yang   Research Center on Fictitious Economy & Data
                                     Science (FEDS),CAS, China
Neil Y. Yen   University of Aizu, Japan
Zhiwen Yu   South China University of Technology, China
Jianke Zhu   Zhejiang University, China
Xiaodong Zhu   University of Shanghai for Science and
                                     Technology, China

BDMA 2013 Workshop Organizers' Message

The First International DASFAA Workshop on Big Data Management and Analytics (BDMA 2013) took place on April 22, 2013, in Wuhan, China, in conjunction with DASFAA 2013, which is an annual international database conference in the Asia-Pacific region. The objective of BDMA 2013 was to create a dedicated forum to bring together researchers, practitioners and others to present and exchange ideas, experiences, and the latest research results on big data management and analytics. BDMA 2013 provided an excellent opportunity for researchers from academia and industry as well as practitioners to showcase the latest advances in this area and to discuss future research directions and challenges on big data management and analytics. The workshop's scope includes processing, management, analytics, visualization, integration and modeling of big data.

We solicit technical papers and position papers (without deep experimental analysis, just to include problem identifications and novel approaches to problem solving) that address all the important aspects of information technologies for processing and analyzing big data. Topics of interest include big data analytics and visualization, big data management architectures, big data placement, scheduling, and optimization, programming models for big data processing, distributed/parallel processing for streaming big data, big data integration and interoperable big data modeling, real-time processing of streaming big data, and streaming big data applications and challenges.

The workshop attracted 10 submissions from France, Korea, China, and the USA. All submissions were peer reviewed by at least two Program Committee members to ensure that high-quality papers were selected. The Program Committee selected nine papers for inclusion in the workshop proceedings.

The Program Committee of the workshop consisted of 21 experienced researchers and experts. We would like to thank the valuable contribution of all the Program Committee members during the peer-review process. We also would like to acknowledge all the authors who submitted very interesting and impressive papers from their work.

April 2013

Cyrus Shahabi
Ge Yu
Hong Bonghee

SNSM 2013 Workshop Organizers' Message

Today the emergence of Web-based communities and hosted services such as social networking sites, wikis and folksonomies brings tremendous freedom of Web autonomy and facilitates collaboration and knowledge sharing between users. Along with the interaction between users and computers, social media are rapidly becoming an important part of our digital experience, ranging from digital textual information to diverse multimedia forms. These aspects and characteristics constitute the core of the second generation of the Web.

A prominent challenge lies in modeling and mining this vast pool of data to extract, represent, and exploit meaningful knowledge and to leverage structures and dynamics of emerging social networks residing in the social media. Social networks and social media mining combines data mining with social computing as a promising direction and offers unique opportunities for developing novel algorithms and tools ranging from text and content mining to link mining.

The 4th International Workshop on Social Network and Social Web Mining (SNSM 2013) was held on April 22, 2013, in Wuhan, China, in conjunction with DASFAA 2013. The overall goal of the workshop was to bring together those in academia, researchers, and industrial practitioners from computer science, information systems, statistics, sociology, behavioral science and organizational science disciplines, and to provide a forum for recent advances in the field of social networks and social media, from the perspectives of data management and mining.

The workshop attracted 12 submissions from France, China, Japan, Korea, and Tunisia. All submissions were peer reviewed by at least three Program Committee members to ensure that high-quality papers were selected. On the basis of the reviews, the Program Committee selected seven papers for inclusion in the workshop proceedings.

The Program Committee of the workshop consisted of 34 experienced researchers and experts. We would like to thank the valuable contribution of all the Program Committee members during the peer-review process. Also, we would like to acknowledge the DASFAA 2013 Workshop Chairs for their great support in ensuring the success of SNSM 2013. Last but not least, we appreciate all the authors who submitted very interesting and impressive papers from their recent work.

April 2013

Guandong Xu
Lin Li

SeCoP 2013 Workshop Organizers' Message

With the ongoing development of the World Wide Web, social networking sites, wikis, and folksonomies are becoming more and more popular and extending the focus of the Semantic Web by engaging it in other communities, where semantics can play an important role. Along with the interaction between users and computers, more and more personalized information is potentially mined from the Web by using semantic computing technology. Connecting semantic computing and personalized computing can enhance classic information management and retrieval approaches. It combines data mining with semantic computing as a promising direction and offers opportunities for developing novel algorithms and tools ranging from text and multimedia.

The International Workshop on Semantic Computing and Personalization (SeCoP) was held in conjunction with DASFAA 2011 on 22 April, 2013. The goal of the workshop is to bring together academics, researchers, and industrial practitioners from computer science, information systems, psychology, behavioral science and organizational science disciplines, and to provide a forum for recent advances in the field of semantic computing and personalization, from the perspectives of information management and mining.

For this first workshop the Program Committee selected three papers for inclusion in the workshop proceedings. We would like to thank the valuable contribution of all the Program Committee members and reviewers. Also, we would like to acknowledge the DASFAA 2013 Conference General Co-chairs and Workshop Co-chairs for their great support in ensuring the success of the SeCop 2013 workshop; we are also grateful for the support we received from the School of Software Engineering at South China University of Technology.

April 2013   Yi Cai

Table of Contents

First International Workshop on Semantic Computing and Personalization (SeCoP)

From Big Data to Big Data Mining: Challenges, Issues, and Opportunities

Dunren Che[1], Mejdl Safran[1], and Zhiyong Peng[2]

[1] Department of Computer Science, Southern Illinois University
Carbondale, Illinois 62901, USA
{mejdl.safran@,dche@cs.}siu.edu
[2] Computer School, Wuhan University, Wuhan, 430072, China
peng@whu.edu.cn

Abstract. While "big data" has become a highlighted buzzword since last year, "big data mining", i.e., mining from big data, has almost immediately followed up as an emerging, interrelated research area. This paper provides an overview of big data mining and discusses the related challenges and the new opportunities. The discussion includes a review of state-of-the-art frameworks and platforms for processing and managing big data as well as the efforts expected on big data mining. We address broad issues related to big data and/or big data mining, and point out opportunities and research topics as they shall duly flesh out. We hope our effort will help reshape the subject area of today's data mining technology toward solving tomorrow's bigger challenges emerging in accordance with big data.

Keywords: data mining, big data, big data mining, big data management, knowledge discovery, data-intensive computation.

1 Introduction

The era of petabyte has come and almost gone, leaving us to confront the exabytes era now. Technology revolution has been facilitating millions of people by generating tremendous data via ever-increased use of a variety of digital devices and especially remote sensors that generate continuous streams of digital data, resulting in what has been called as "big data". It has been a confirmed phenomenon that enormous amounts of data have been being continually generated at unprecedented and ever increasing scales. In 2010, Google estimated that every two days at that time the world generated as much data as the sum it generated up to 2003. Regardless of the very recent "Big Data Executive Survey 2013" by NewVantage Partners [36] that states "It's about variety, not volume", many people (including the authors) would still believe the foremost issue with big data is scale or *volume*. Big data sure involves a great *variety* of data forms: text, images, videos, sounds, and whatever that may come into the play, and their arbitrary combinations (the type system shall remain constantly open). Big data frequently comes in the form of streams of a variety of

B. Hong et al. (Eds.): DASFAA Workshops 2013, LNCS 7827, pp. 1–15, 2013.

types. Time is an integral dimension of data streams, which often implies that the data must be processed/mined in a timely or (nearly) real-time manner. Besides, the current major consumers of big data, corporate businesses, are especially interested in "a big data environment that can accelerate the time-to-answer critical business questions that demonstrate business values" [36]. The time dimension of bid data naturally leads to yet another key characteristic of big data – speed or *velocity*. We concur that big data is not *all* about size or volume, but size is the foremost characteristic of big data – size sparks off a series of interrelated vital challenges beyond size itself. Gartner analysts [29, 3] described the dominant characteristics of big data as "three Vs" – Volume, Velocity, and Variety (alternatively referred to as V3). Serious challenges are unfolded along each of the "V" axis. Bearing the brunt of criticism, the once very successful DBMSs have denounced no longer being able to meet the increasing demands of big data – it was "too big, too fast, and too hard" for existing DBMSs and tools [4] to satisfactorily handle. These challenges call for a new stack (or many alternate stacks) of highly scalable computing models, tools, frameworks and platforms, etc., being capable to tap into the most potential of today's best parallel and elastic computing facility – cloud computing.

Scalability is at the core of the expected new technologies to meet the challenges coming along with big data. The simultaneously emerging and fast maturing cloud computing technology delivers the most promising platforms to realize the needed scalability with demonstrated elasticity and parallelism capacities. Numerous notable attempts have been initiated to exploit massive parallel processing architectures as reported in [5], [6], [7], [8], [9], [26], [19] and [17]. Google's novel programing model, MapReduce [5], and its distributed file system, GFS (Google File System) [6], represent the early groundbreaking efforts made in this line. We will shed more lights on these representative works later in this paper.

From the data mining perspective, mining big data has opened many new challenges and opportunities. Even though big data bears greater value (i.e., hidden knowledge and more valuable insights), it brings tremendous challenges to extract these hidden knowledge and insights from big data since the established process of knowledge discovering and data mining from conventional datasets was not designed to and will not work well with big data. The cons of current data mining techniques when applied to big data are centered on their inadequate scalability and parallelism. In general, existing data mining techniques encounter great difficulties when they are required to handle the unprecedented heterogeneity, volume, speed, privacy, accuracy, and trust coming along with big data and big data mining. Improving existing techniques by applying massive parallel processing architectures and novel distributed storage systems, and designing innovative mining techniques based on new frameworks/platforms with the potential to successfully overcome the aforementioned challenges will change and reshape the future of the data mining technology. Numerous research projects, as reported in [11], [12], [13], [14], [15] and [16], have been initiated in the last couple of years for the sake of overcoming the big data challenges. We will shed more lights on these projects later in this paper.

The theme of this paper is to provide an in-depth study on the issue of big data mining, its challenges and the perceivable opportunities. We also point to a few research topics that are either promising or much needed for solving the big data and big data mining problems. In order to make our discussion logical and smooth, we will start with a review of some essential and relevant concepts, including data mining, big data, big data mining, and the frameworks/platforms (completed or under construction) related to big data and big data mining.

The remainder of this paper is organized as follows. In Section 2, we briefly review the data mining concept and the foreseen challenges when the technology is applied to big data. In Section 3, we examine the concept of big data, comparing with conventional databases and reviewing the emerging platforms designed for big data and big data mining. In Section 4, we revisit data mining in the new context of big data, and point out emerging challenges. In Section 5, we discuss additional issues and challenges related to big data mining (in this section we especially draw attention to privacy crisis and garbage mining, which do not seem have been addressed by anyone else, to the best of our knowledge). Section 6 concludes our discussion.

2   Data Mining

Knowledge discovery (KDD) is a process of unveiling hidden knowledge and insights from a large volume of data [1], which involves data mining as its core and the most challenging and interesting step (while other steps are also indispensable) . Typically, data mining uncovers interesting patterns and relationships hidden in a large volume of raw data, and the results tapped out may help make valuable predictions or future observations in the real world. Data mining has been used by a wide range of applications such as business, medicine, science and engineering. It has led to numerous beneficial services to many walks of real businesses – both the providers and ultimately the consumers of services.

Applying existing data mining algorithms and techniques to real-world problems has been recently running into many challenges due to the inadequate scalability (and other limitations) of these algorithms and techniques that do not match the three Vs of the emerging big data. Not only the scale of data generated today is unprecedented, the produced data is often continuously generated in the form of streams that require being processed and mined in (nearly) real time. Delayed discovery of even highly valuable knowledge invalidates the usefulness of the discovered knowledge. Big data not only brings new challenges, but also brings opportunities – the interconnected big data with complex and heterogeneous contents bear new sources of knowledge and insights. Big data would become a useless monster if we don't have the right tools to harness its "wildness". We argue to consider big data as greatly expanded assets to human. All what we need then is to develop the right tools for efficient *store*, *access*, and *analytics* (SA2 for short). Current data mining techniques and algorithms are not ready to meet the new challenges of big data. Mining big data demands highly scalable strategies and algorithms, more effective preprocessing steps such as data filtering and integration, advanced parallel computing environments (e.g., cloud Paas and

IaaS), and intelligent and effective user interaction. Next we examine the concept and big data and related issues, including emerging challenges and the (foregoing and ongoing) attempts initiated on dealing with big data.

3   Big Data

We are sure living in an interesting era – the era of big data and cloud computing, full of challenges and opportunities. Organizations have already started to deal with petabyte-scale collections of data; and they are about to face the exabyte scale of big data and the accompanying benefits and challenges.

Big data is believed to play a critical role in the future in all walks of our lives and our societies. For example, governments have now started mining the contents of social media networks and blogs, and online-transactions and other sources of information to identify the need for government facilities, to recognize the suspicious organizational groups, and to predict future events (threats or promises). Additionally, service providers start to track their customers' purchases made through online, in-store, and on-phone, and customers' behaviors through recorded streams of online-clicks, as well as product reviews and ranking, for improving their marketing efforts, predicting new growth points of profits, and increasing customer satisfaction.

The mismatch between the demands of the big data management and the capabilities that current DBMSs can provide has reached the historically high peak. The three Vs (volume, variety, and velocity) of big data each implies one distinct aspect of critical deficiencies of today's DBMSs. Gigantic volume requires equally great scalability and massive parallelism that are beyond the capability of today's DBMSs; the great variety of data types of big data particularly unfits the restriction of the closed processing architecture of current database systems [4]; the speed/velocity request of big data (especially stream data) processing asks for commensurate real-time efficiency which again is far beyond where current DBMSs could reach. The limited *availability* of current DBMSs defeats the velocity request of big data from yet another angle (Current DBMSs typically require to first import/load data into their storage systems that enforces a uniform format before any access/processing is allowed. Confronted with the huge volume of big data, the importing/loading stage could take hours, days, or even months. This causes substantially delayed/reduced availability of the DBMSs).

To overcome this scalability challenge of big data, several attempts have been made on exploiting massive parallel processing architectures. The first such attempt was made by Google. Google created a programming model named MapReduce [5] that was coupled with (and facilitated by) the GFS (Google File System [6]), a distributed file system where the data can be easily partitioned over thousands of nodes in a cluster. Later, Yahoo and other big companies created an Apache open-source version of Google's MapReduce framework, called Hadoop MapReduce. It uses the Hadoop Distributed File System (HDFS) – an open source version of the Google's GFS. The MapReduce framework allows users to define two functions, map and reduce, to process a large number data entries in parallel [7]. More specifically, in MapReduce, the input is divided into a large set of key-value pairs first; then the map function is called and forked into many instances concurrently processing on the large key-value pairs. After all data entries are processed, a new set of key-value pairs are produced, and then the reduce function is called to group/merge the produced values based on common keys.

In order to match/support the MapReduce computing model, Google developed the BigTable – a distributed storage system designed for managing structured data. BigTable can scale well to a very large size: petabytes of data across thousands of commodity severs [8]. In the same spirit, Amazon created Dynamo [9], which is also a key-value pair storage system. The Apache open-source community acted quickly again, created HBase – an open-source version of Google's BigTable built on top of HDFS and Cassandra – an open-source version of Amazon's Dynamo. Apache Hive [25] is an open source data warehouse system built on top of Hadoop for querying and analyzing files stored in HDFS using a simple query language called HiveQL.

Hadoop is not alone; it has other competitor platforms. All these platforms lack many niceties existing in DBMSs. Some of the competitors improved on existing platforms (mostly on Hadoop), and others came up with a fresh system design. However, most of these platforms are still in their infancy. For example, BDAS, the Berkeley Data Analytics Stack [26], is an open-source data analytics stack developed at the UC Berkeley AMPLab for computing and analyzing complex data. It includes the following main components: Spark, Shark,, and Mesos. Spark is a high-speed cluster computing system that performs computations in memory and can outperform Hadoop by up to 100x. Shark is a large-scale data analysis system for Spark that provides a unified engine running SQL queries, compatible with Apache Hive. Shark can answer SQL queries up to 100x faster than Hive, and run iterative machine learning algorithms up to 100x faster than Hadoop, and can recover from failed mid-queries within seconds [27]. Mesos is a cluster manager that can run Hadoop, Spark and other frameworks on a dynamically shared pool of compute nodes. ASTERIX [19] is data-intensive storage and computing platform. As a research project, it was initiated by "the three database guys" at UC Irvine. Some notable drawbacks of Hadoop and other similar platforms, e.g., single system performance, difficulties of future maintenance, inefficiency in pulling data up to queries and the unawareness of record boundaries, are properly overcome in ASTERIX by exploring runtime models inspired by parallel database system execution engines [19]. In ASTERIX, the open software stack is layered in a different way that it sets the data records at the bottom layer, facilitating a higher-level language API at the top. While the majority of the big data management and processing platforms have been (or are being) developed to meet business needs, SciDB [17] is an open source data management and analytics (DMAS) software system for data-intensive scientific applications like radio astronomy, earth remote sensing and environment observation and modeling. The difference between SciDB and other platforms is that SciDB is designed based on the concept of array DBMS (i.e., raster data) where big data is represented as arrays of objects in unidimensional or multidimensional spaces. SciDB is designed to support integration with high-level imperative languages, algorithms, and very large scales of data [4].

In the next section, we discuss the attempts related to big data mining.

4   Big Data Mining

The goals of big data mining techniques go beyond fetching the requested information or even uncovering some hidden relationships and patterns between numeral parameters. Analyzing fast and massive stream data may lead to new valuable insights and theoretical concepts [2]. Comparing with the results derived from mining the conventional datasets, unveiling the huge volume of interconnected heterogeneous big data has the potential to maximize our knowledge and insights in the target domain. However, this brings a series of new challenges to the research community. Overcoming the challenges will reshape the future of the data mining technology, resulting in a spectrum of groundbreaking data and mining techniques and algorithms. One feasible approach is to improve existing techniques and algorithms by exploiting massively parallel computing architectures (cloud platforms in our mind). Big data mining must deal with heterogeneity, extreme scale, velocity, privacy, accuracy, trust, and interactiveness that existing mining techniques and algorithms are incapable of.

The need for designing and implementing very-large-scale parallel machine learning and data mining algorithms (ML-DM) has remarkably increased, which accompanies the emergence of powerful parallel and very-large-scale data processing platforms, e.g., Hadoop MapReduce. NIMBLE [11] is a portable infrastructure that has been specifically designed to enable rapid implementation of parallel ML-DM algorithms, running on top of Hadoop. Apache's Mahout [12] is a library of machine learning and data mining implementations. The library is also implemented on top of Hadoop using the MapReduce programming model. Some important components of the library can run stand-alone. The main drawbacks of Mahout are that its learning cycle is too long and its lack of user-friendly interaction support. Besides, it does not implement all the needed data mining and machine learning algorithms. BC-PDM (Big Cloud-Parallel Data Mining) [13], as a cloud-based data mining platform, also based on Hadoop, provides access to large telecom data and business solutions for telecom operators; it supports parallel ETL process (extract, transform, and load), data mining, social network analysis, and text mining. BC-PDM tried to overcome the problem of single function of other approaches and to be more applicable for Business Intelligence. PEGASUS (Peta-scale Graph Mining System) [14] and Giraph [15] both implement graph mining algorithms using parallel computing and they both run on top of Hadoop. GraphLab [16] is a graph-based, scalable framework, on which several graph-based machine learning and data mining algorithms are implemented. The reported drawback of GraphLab is that it requires all data fitting into memory.

In the next section, we further expand our discussion along the key issues and challenges of big data mining.

5   Issues and Challenges

Our subsequent discussion centers on the following key issues and challenges: heterogeneity (or variety), scale (or volume), speed (or velocity), accuracy and trust, privacy crisis, interactiveness, and garbage mining (This section is supposedly the most interesting one of this paper).

5.1　Variety and Heterogeneity

In the past, data mining techniques have been used to discover unknown patterns and relationships of interest from structured, homogeneous, and small datasets (from today's perspective). Variety, as one of the essential characteristics of big data, is resulted from the phenomenon that there exists nearly unlimited different sources that generate or contribute to big data. This phenomenon naturally leads to the great variety or heterogeneity of big data. The data from different sources inherently possesses a great many different types and representation forms, and is greatly interconnected, interrelated, and delicately and inconsistently represented. Mining such a dataset, the great challenge is perceivable and the degree of complexity is not even imaginable before we deeply get there. Heterogeneity in big data also means that it is an obligation (rather than an option) to accept and deal with structured, semi-structured, and even entirely unstructured data simultaneously. While structured data can fit well into today's database systems, semi-structured data may partially fit in, but unstructured data definitely will not. Both semi-structured and unstructured data are typically stored in files. This is especially so in data-intensive, scientific computation areas [37]. Nevertheless, though bringing up greater technical challenges, the heterogeneity feature of big data means a new opportunity of unveiling, previously impossible, hidden patterns or knowledge dwelt at the intersections within heterogeneous big data. We shed a little more light on the implied challenge and the opportunity by looking into the examples from a familiar scenario in the following.

First, as a classic data mining example, we consider a simple grocery transaction dataset that records only one type of data, i.e., goods items. Examples insights [10] that might be mined from this dataset may include, e.g., the famous association of "beer and diapers" showing a strong linkage between the two items, and popular items like milk that are almost always purchased by customers, showing strong linkage of milk to all other items. In contrast to that, big data mining must deal with semi-structured and heterogeneous data. Now we generalize the aforementioned simple example by extending the scenario to an online market such as eBay. The dataset now is a richer network consisting of at least three different types of objects: items, buyers, and sellers (still this scenario may not be considered complex enough to demonstrate the complexity in big data mining). Interrelation may broadly exist, e.g., between commodity items in the form of "bought with", between sellers and items in the form of "sell" and "sold by", between buyers and items in the form of "buy" or "bought by", and between buyers and sellers in the form of "buy from" and "sold to". This data network has different types of objects and relationships (indicating a light shade of heterogeneity). We speculate that existing data mining techniques would not (if applicable at all) maximally uncover the hidden associations and insights in this data network.

For a heterogeneous set of big data, trying to construct a single model (if doable at all) would most likely not result in good-enough mining results; thus constructing specialized, more complex, multi-model systems is expected [21]. An interesting algorithm following this spirit is proposed in [22] that first determines whether the given dataset is truly heterogeneous, and if so, it then partitions the set into homogeneous subsets and constructs a specialized model for each homogeneous subset. Partitioning, as an intuitive approach, would speed up the process of knowledge discovery from heterogeneous big data. However, potential patterns and knowledge may miss the opportunity of being discovered after partitioning if important relationships (often implicit) crossing distinct homogeneous regions are not adequately retained.

The social community mining problem has recently received a lot attention from the researchers. This problem desires "multi-network, user-dependent, and query-based analysis" [24]. It conveys that the intersections between multiple networks bear potential knowledge and insights that may not be discovered if a homogenous model is to be enforced.

Mining from heterogeneous information networks is a promising frontier of current data mining research [20]. Relational databases have been used to capture the heterogeneous information networks and new methods for in-depth network-oriented data mining and analysis have been proposed [20]. However, the degree of the heterogeneity captured does not reflect the real degree of the inherent heterogeneity existing in the big data. Mining hidden patterns from heterogeneous multimedia streams of diverse sources represents another frontier of data mining research. The output of this research has broad applicability such as detection of spreading dangerous diseases and prediction of traffic patterns and other critical social events (e.g., emerging conflicts and wars).

Like data mining, the process of big data mining shall also starts with data selection (from multiple sources). Data filtering, cleaning, reduction, and transformation then follow. There emerge new challenges with each of these preprocessing steps. With data filtering, how do we make sure that the discarded data will not severely degrade the quality of the eventually mined results under the complexity of great heterogeneity of big data? The same question could be adapted and asked to all other preprocessing steps and operations of the data mining process.

5.2   Scalability

The unprecedented volume/scale of big data requires commensurately high scalability of its data management and mining tools. Instead of being timid, we shall proclaim the extreme scale of big data because more data bears more potential insights and knowledge that we have no chance to discover from conventional data (of smaller scales). We are optimistic with the following approaches that, if exploited properly, may lead to remarkable scalability required for future data and mining systems to manage and mine the big data: (1) cloud computing that has already demonstrated admirable elasticity, which, combined with massively parallel computing architectures, bears the hope of realizing the needed scalability for dealing with the volume challenge of big data; (2) advanced user interaction support (either GUI- or language-based) that facilitates prompt and effective system-user interaction. Big data mining straightforwardly implies extremely time-consuming navigation in a gigantic search space, and prompt feedback/interference/guidance from users (ideally domain experts) must be beneficially exploited to help make early decisions, adjust search/mining strategies on the fly, and narrow down to smaller but promising sub-spaces.

5.3   Speed/Velocity

For big data, speed/velocity really matters. The capability of fast accessing and mining big data is not just a subjective desire, it is an obligation especially for data streams (a common format of big data) – we must finish a processing/mining task within a certain period of time, otherwise, the processing/mining results becomes less valuable or even worthless. Exemplary applications with real-time requests include earthquake prediction, stock market prediction and agent-based autonomous exchange (buying/selling) systems. Speed is also relevant to scalability – conquering or partially solving anyone helps the other one.

The speed of data mining depends on two major factors: data access time (determined mainly by the underlying data system) and, of course, the efficiency of the mining algorithms themselves. Exploitation of advanced indexing schemes is the key to the speed issue. Multidimensional index structures are especially useful for big data. For example, a combination of R-Tree and KD-tree [30] and the more recently proposed FastBit [18, 23] (developed by the data group at LBNL) shall be considered for big data. Besides, design of new and more efficient indexing schemes is much desired, but remains one of the greatest challenges to the research community.

An additional approach to boost the speed of big data access and mining is through maximally identifying and exploiting the potential parallelism in the access and mining algorithms. The elasticity and parallelism support of cloud computing are the most promising facilities for boosting the performance and scalability of big data mining systems. It is interesting to note that the MapReduce parallel computing model is applicable to only a rather limited class of data-intensive computing problems. Therefore, design of new and more efficient parallel computing models besides MapReduce is greatly desired, but calls for really creative minds.

5.4   Accuracy, Trust, and Provenance

In the past, data mining systems were typically fed with relatively accurate data from well-known and quite limited sources, so the mining results tend to be accurate, too; thus accuracy and trust have never been a serious issue for concern. With the emerging big data, the data sources are of many different origins, not all well-known, and not all verifiable. Therefore, the accuracy and trust of the source data quickly become an issue, which further propagates to the mining results as well. To (at least partially) solve this problem, data validation and provenance tracing become more than a necessary step in the whole knowledge discovery process (including data mining). History has repeatedly proven that challenges always comes hand-in-hand with opportunities (sometimes unnoticeably). In the case of big data, the copious data sources and gigantic volumes provide rich sources to extract additional evidences for verifying accuracy and building trust on the selected data and the produced mining results.

The vast volume of big data attributes additional characteristics – high dynamics and evolution. So an adequate system for big data management and analysis must allow dynamic changing and evolution of the hosted data items. This makes *data provenance* an integral feature in any system that deals with big data [28]. Provenance relates to the evolution history or the origin that a data item was extracted or collected from. The provenance relationships in big data often form a large collection of interrelated derivation chains, resulting in, more generally, a DAG. Trust measures are not and should not be treated static. When data evolves, trust measures shall change or be updated, too. Several unsupervised learning methods have been proposed in [31] and [32] to discover the trust measures of suspected data sources using other data sources as testimony (Here the assumed philosophy of proof is that one does not adequately prove himself innocent without having a third party's testimony). Reference [33] has shown that semi-supervised learning methods that start with ground truth data may provide higher accuracy and trust on the source data. In the context of big data, innovative methods that can run on parallel platforms (such as cloud PaaS and IaaS) dealing with scalable data with numerous sources are highly desired.

Provenance directly contributes to accuracy and trust of the source data and the derived (or mined) results. However, provenance information may not be always recorded or available. When the missing provenance of some data becomes a keen interest of the users, data mining can be reversely applied to derive and verify the provenance. Without a great many sources in the past, many provenance mining problems are unsolvable. History and archeology researches have raised a very interesting class of provenance mining problems. For example, the old question that whether Native Americans were originated from eastern Asia, after decades of debates, is still undetermined. With the advent of big data and mining tools, now we can glimpse the hope of finding the best answer to this and other questions of this type in the near future. We would rather believe the World Wide Web, as the largest data and knowledge base (indeed the Google executives firmly hold on this vision), bears sufficient information needed to derive the best answer to this and other similar questions, and yet the volume of this largest big data repository still keeps growing at an unprecedented pace. We foresee the big data mining technology will soon be able to answer many big questions like the above one though mining the whole World Wide Web as a single dataset (Digesting, consolidating, and deriving the best answer to the above question require the capacity that is way beyond the human brainpower).

5.5   Privacy Crisis

Data privacy has been always an issue even from the beginning when data mining was applied to real-world data. The concern has become extremely serious with big data mining that often requires personal information in order to produce relevant/accurate results such as location-based and personalized services, e.g., targeted and individualized advertisements. Also, with the huge volume of big data such as social media that contains tremendous amount of highly interconnected personal information, every piece of information about everybody can be mined out, and when all pieces of the information about a person are dug out and put together, any privacy about that individual instantly disappears. You might ask, how could this be possible? Well, it is already a reality that every transaction regarding our daily life is being pushed to online and leaves a trace there: we comminute with friends via email, instant message, blog, and Facebook; we do shopping and pay our bills online too; and yet, credit card companies hold our confidential identity information; your payroll office has your personal information, too; your home phone number and address are listed in the region's directory that everyone can access; last month, you had a birthday party that disclosed your exact birthday to the circle of your friends, and some of them posted your birthday party in blogs, ... Thanks goodness, everyone so far has the righteous sense of protecting your confidential personal information, but the possibility of unintended leaking cannot be ruled out once and forever, and no leaking today does not guarantee impermeable tomorrow. As time goes, every piece of your personal information will be scattered here or there (hopefully not all available from one location). Well, we have desperately wanted and are diligently working toward powerful mining tools capable of mining a great portion or even the whole Web. So you shall not doubt such powerful mining tools or systems one day will be able to find confidential information of you (and actually of everyone else) – it's now just a matter of time. Everyone would easily gain the privilege of using such powerful tools (via SaaS on the cloud), mine your privacy, and see you entirely "naked". Without the shield of any privacy protecting you, a bad guy could open a new credit card account in your name, and transfer your hard-earned money away from your bank account... Everything seems becoming possible! Imagine how big a social disaster it would be when everyone in the US, for example, can access everyone else's social security number and other identity information, name, address, birthday, birthplace, phone numbers, etc. Even credit card companies do not ask for all this information when one requests to open a new account on the phone. So we definitely run the risk of living transparently or "naked" in an era of no privacy. Should we be proud to say that one day, we will live in a world that everyone can perfectly pretend to be any other one? Well, when anybody can "become" another body as s/he wishes, we get completely separated from our true identities. Now we need most seriously ask ourselves: would we rather to wear the "the emperor's new clothes"? The answer is certainly "no" as we all believe. Then what are the possible countermeasures? Apparently, we urgently need proper policies and approaches to manage sharing of personal data, while legitimate data mining activities shall still be granted facilitated. As said in [34], the privacy issue calls for "the development of a model where the benefits of data for businesses and researchers are balanced against individual privacy rights" [34]. The foundations of data mining need to be reformulated when dealing with big data "in such a way that privacy protection and discrimination prevention are embedded in the foundations themselves, dealing with every moment in the data-knowledge life-cycle: from (off-line and on-line) data capture, to data mining and analytics, up to the deployment of the extracted models" [35]. Measuring and prevention of privacy violation during knowledge mining are two related issues that call for serious research and innovative solutions.

5.6   Interactiveness

By interactiveness we mean the capability or feature of a data mining system that allows prompt and adequate user interaction such as feedback/interference/guidance from users. Interactiveness is relatively an underemphasized issue of data mining in the past. When our society is now confronting the challenges of big data mining, interactiveness becomes a critical issue. Interactiveness relates to all the "three Vs" and can help overcome the challenges coming along with each of them. First, as we pointed out earlier, in order to conquer the volume related challenge of big data mining, prompt user feedback/guidance can help quickly narrow down into a much reduced but promising sub-space, accelerate the processing speed (or velocity) and increase system scalability. Second, the heterogeneity caused by the variety of big data straightforwardly induces accordingly high complexity in the big data itself and the mining results. Sufficient system interactiveness grants users the ability to visualize, (pre-)evaluate, and interpret intermediate and final mining results. Such a facility might not be quite necessary for mining conventional datasets, but for big data, it is a must.

Great interactiveness boosts the acceptance of a complicated mining system and its mining results by potential users. In short, the head of the pyramid would be missing if adequate user interaction is not supported. Even though a data mining system has been very professionally designed, with perfect functional layers, without adequate interactiveness, the value of the system would be greatly discounted or simply rejected by users. Sufficient interactiveness is especially important for big data mining.

5.7   Garbage Mining

Who wants garbage when there are potentially gold? Garbage has no value. No one wants garbage. Everyone wants to get rid of garbage. In the real world, garbage collection is a business with profits. Garbage does not speak: "I am garbage, recycle me!" At home, our rooms are filled with stuff, and many items may never be needed, but we lack the wisdom to realize for sure. We easily fill up a 1000 GB disk in our desktop computers, whereas, only a small portion hoarded there are useful files (most of us would wholeheartedly agree on this!). We are not willing to spend time to clean up our disk space, more often, our memory becomes blurry as time goes and we don't remember the difference between two seemingly identical data files, and which file holds important consolidated data copied from other files that shall thus be recycled but we just did not promptly do so. Even cleaning up the disk space of desktop computer is a headache, not to mention to clean up the cyberspace! It has been a common sight that, e.g., you were searching the internet for customers' reviews and recommendations, say, for a good air-conditioning servicer in your area, and a professionally written blog caught your eyes, commending someone that you found already moved off the region after you made a couple of phone calls, and then you glimpsed the blog again, realizing the post date was in 2004. The blog space should have been cleaned; outdated and meaningless comments should have been deleted. Unfortunately, this phenomenon does not only occur with blogs, it is common with the entire cyberspace. In the big data era, the volume of data generated and populated on the World Wide Web keeps increasing at an amazingly fast pace. In such an environment, data can (quickly) become outdated, corrupted, and useless; in addition, there is data that is created as junks (like junk emails). If the society does not pay attention and take actions now, as time goes, we will be flooded by junk data in the cyberspace. For the sake of having a relatively clean cyberspace and clean World Wide Web, herein we call for attentions and research efforts. Cyberspace cleaning is not an easy task because of at least two foreseeable reasons: garbage is hidden, and there is an ownership issue – are you granted to collect someone else's garbage (provided you have the motivation)?

We propose applying data mining approaches to mine garbage and recycle it. We haven't yet noticed (to the best of our knowledge) the issue being realized and discussed anywhere else. But we believe garbage mining is a serious research topic, different but related to big data mining – for the sake the *sustainability* of our digital environment, "mining for garbage" (and cleaning it) is as important as "mining for knowledge" (the canonical sense of data mining). This is especially so in the new era of big data.

We envision that in the future the society will develop mobile intelligent scavenger agents (with embedded garbage mining modules) and dispatch them to the cyberspace to autonomously and legitimately mine and clean up garbage in the cyberspace. Similarly, local versions of the intelligent scavenger agents shall be created and used to help clean up the disk space of desktop computers, if not entirely autonomously, at least interactively with necessary guidance and confirmation prompted from the users.

"One man's trash is another's treasure". Garbage definition remains one of the greatest challenges.

6   Conclusion

We are living in the big data era where enormous amounts of heterogeneous, semi-structured and unstructured data are continually generated at unprecedented scale. Big data discloses the limitations of existing data mining techniques, resulted in a series of new challenges related to big data mining. Big data mining is a promising research area, still in its infancy. In spite of the limited work done on big data mining so far, we believe that much work is required to overcome its challenges related to heterogeneity, scalability, speed, accuracy, trust, provenance, privacy, and interactiveness. This paper also provides an overview (though limited due to space limit) of state-of-the-art frameworks/platforms for processing and managing big data as well as platforms and libraries for mining big data. More specifically, we originally pointed out and analyzed the risk of privacy crisis which is deteriorated by big data and big data mining (Section 5.5) and first time proposed and formulated garbage mining – a critical issue in the big data era that has not been realized by others nor addressed anywhere else (Section 5.7). As our future work, we are at the stage of seriously planning a research project on cyberspace garbage mining to make the cyberspace a more sustainable environment. We tried to fill our discussions with sparking, constructive ideas. We hope we have (at least partially) gotten there.

References

1. Fayyad, U.M., Gregory, P.S., Padhraic, S.: From Data Mining to Knowledge Discovery: an Overview. In: Advances in Knowledge Discovery and Data Mining, pp. 1–36. AAAI Press, Menlo Park (1996)
2. Berkovich, S., Liao, D.: On Clusterization of big data Streams. In: 3rd International Conference on Computing for Geospatial Research and Applications, article no. 26. ACM Press, New York (2012)
3. Beyer, M.A., Laney, D.: The Importance of 'Big Data': A Definition. Gartner (2012)
4. Madden, S.: From Databases to big data. IEEE Internet Computing 16(3), 4–6 (2012)
5. Dean, J., Ghemawat, S.: MapReduce: Simplified Data Processing on Large Clusters. In: 6th Symposium on Operating System Design and Implementation (OSDI), pp. 137–150 (2004)
6. Ghemawat, S., Gobioff, H., Leung, S.T.: The Google File System. In: 19th ACM Symposium on Operating Systems Principles, Bolton Landing, New York, pp. 29–43 (2003)
7. Dean, J., Ghemawat, S.: MapReduce: a Flexible Data Processing Tool. Communication of the ACM 53(1), 72–77 (2010)
8. Chang, F., Dean, J., Ghemawat, S., et al.: Bigtable: A Distributed Storage System for Structured Data. In: 7th Symposium on Operating Systems Design and Implementation, vol. 7, pp. 205–218. USENIX Association Berkeley, CA (2006)
9. DeCandia, G., Hastorun, D.: Jampani, et al: Dynamo: Amazon's Highly Available Key-Value Store. In: 21st ACM SIGOPS Symposium on Operating Systems Principles, pp. 14–17. Stevenson, Washington (2007)
10. Shmueli, G., Patel, N.R., Bruce, P.C.: Data Mining for Business Intelligence: Concepts, Techniques, and Applications in Microsoft Office Excel with XLMiner, 2nd edn. Wiley & Sons, Hoboken (2010)
11. Ghoting, A., Kambadur, P., Pednault, E., Kannan, R.: NIMBLE: a Toolkit for the Implementation of Parallel Data Mining and Machine Learning Algorithms on MapReduce. In: 17th ACM SIGKDD International Conference on Knowledge Discovery and Data Mining, San Diego, California, USA, pp. 334–342 (2011)
12. Mahout, http://lucene.apache.org/mahout/
13. Yu, L., Zheng, J., Shen, W.C., et al.: BC-PDM: Data Mining, Social Network Analysis and Text Mining System Based on Cloud Computing. In: 18th ACM SIGKDD International Conference on Knowledge Discovery and Data Mining, pp. 1496–1499 (2012)
14. Kang, U., Tsourakakis, C.E., Faloutsos, C.: PEGASUS: A Peta-Scale Graph Mining System Implementation and Observations. In: 9th IEEE International Conference on Data Mining, pp. 229–238 (2009)
15. Apache Giraph Project, http://giraph.apache.org/
16. Low, Y., Bickson, D., Gonzalez, J., Guestrin, C., Kyrola, A., Hellerstein, J.M.: Distributed GraphLab: A Framework for Machine Learning and Data Mining in the Cloud. VLDB Endowment 5(8), 71–727 (2012)
17. Brown, P.G.: Overview of SciDB: Large Scale Array Storage, Processing and Analysis. In: ACM SIGMOD International Conference on Management of Data, pp. 963–968 (2010)
18. Wu, K.: FastBit: An Efficient Indexing Technology for Accelerating Data-intensive Science. Journal of Physics, Conference Series 16, 550–560 (2005)
19. Borkar, V.R., Carey, M.J., Li, C.: big data Platforms: What's Next? ACM Crossroads 19(1), 44–49 (2012)

20. Sun, Y., Han, J., Yan, X., Yu, P.S.: Mining Knowledge from Interconnected Data: A Heterogeneous Information Network Analysis Approach. VLDB Endowment 5(12), 2022–2023 (2012)
21. Obradovic, Z., Vucetic, S.: Challenges in Scientific Data Mining: Heterogeneous, Biased, and Large Samples. Technical Report, Center for Information Science and Technology Temple University, ch. 1, pp. 1–24 (2004)
22. Vucetic, S., Obradovic, Z.: Discovering Homogeneous Regions in Spatial Data through Competition. In: 17th International Conference of Machine Learning, Stanford, CA, pp. 1095–1102 (2000)
23. Wu, K., Ahern, S.: Bethel, et al: FastBit: Interactively Searching Massive Data. SciDAC 180 (2009)
24. Cai, D., Shao, Z., He, X., Yan, X., Han, J.: Mining Hidden Communities in Heterogeneous Social Network. In: 3rd International Workshop Link Discovery (LinkKDD), pp. 58–65 (2005)
25. Apache Hive, http://hive.apache.org/
26. Berkeley Data Analytics Stack (BDAS), https://amplab.cs.berkeley.edu/bdas/
27. Xin, R.S., Rosen, J., Zaharia, M., Franklin, M., Shenker, S., Stoica, I.: Shark: SQL and Rich Analytics at Scale. In: ACM SIGMOD Conference (accepted, 2013)
28. Agrawal, D., Bernstein, P., Bertino, E., et al.: Challenges and Opportunities With big data – A Community White Paper Developed by Leading Researchers Across the United States (2012), http://cra.org/ccc/docs/init/bigdatawhitepaper.pdf
29. Laney, D.: 3D Data Management: Controlling Data Volume, Velocity and Variety. Gartner (2001)
30. Zhang, X., Ai, J., Wang, Z., Lu, J., Meng, X.: An Efficient Multi-dimensional Index for Cloud Data Management. In: 1st International Workshop on Cloud Data Management, pp. 17–24. ACM Press, Hong Kong (2009)
31. Yin, X., Han, J., Yu, P.S.: Truth Discovery with Multiple Conflicting Information Providers on the Web. In: 13th ACM SIGKDD International Conference on Knowledge Discovery and Data Mining, San Jose, California, pp. 1048–1052 (2007)
32. Dong, X.L., Berti-Equille, L., Srivastava, D.: Integrating Conflicting Data: The Role of Source Dependence. VLDB Endowment 2(1), 550–561 (2009)
33. Yin, X., Tan, W.: Semi-Supervised Truth Discovery. In: 20th International Conference on World Wide Web, Hyderabad, India, pp. 217–226 (2011)
34. Tene, O., Polonetsky, J.: Privacy in the Age of big data: A Time for Big Decisions. Stanford Law Review Online 64, 63–69 (2012)
35. Pedreschi, D., Calders, T., Custers, B., et al.: big data Mining, Fairness and Privacy - A Vision Statement Towards an Interdisciplinary Roadmap of Research. Data Mining and Analytics Software, KDnuggets Review Online 11(26) (2011)
36. NewVantage Partners: Big Data Executive Survey (2013), http://newvantage.com/wp-content/uploads/2013/02/NVP-Big-Data-Survey-2013-Summary-Report.pdf
37. Greenwald, M., Fredian, T., Schissel, D., Stillerman, J.: A Metadata Catalog for Organization and Systemization of Fusion Simulation Data. Fusion Engineering & Design 87(12), 2205–2208 (2012)

An Economical Query Cost Model in the Cloud

Assia Brighen[1], Ladjel Bellatreche[2], Hachem Slimani[1], and Zoé Faget[2]

[1] Computer Science Department
University of Bejaia, Algeria
abrighen@yahoo.fr, haslimani@gmail.com
[2] LIAS/ISAE-ENSMA - University of Poitiers
Futuroscope 86960, Cedex France
{bellatreche,zoe.faget}@ensma.fr

Abstract. The Cloud Computing ($\mathcal{CC}$) brings a new approach of information technology (IT) consumption and is changing the investment manner of enterprises and companies. While the reputation of $\mathcal{CC}$ is increasing, a large number of applications managing large amount of data are moving towards the Cloud which incorporates several new dimension: payment, query processing, etc.. The analytical data management applications are an example of those applications. They are intended to the decision support process requiring complex queries. To optimize these queries, optimization structures such as indexes and materialized views are required. In the traditional database infrastructures (centralized, parallel, distributed, etc.), the choice of the optimal configuration of optimization structures is usually guided by mathematical cost models. They are used to quantify the quality of the obtained solutions. The purpose of our work is to develop a cost model to select materialized views in the Cloud. The main characteristic of our cost model is that it considers the payment cost and the query processing paradigm. Intensive experiments were conducted using our cost model and the obtained results are deployed in an assimilated Cloud infrastructure.

1 Introduction

Cloud Computing ($\mathcal{CC}$) is a new paradigm that refers to the use of computing resources on demand through a global network. It defines a new way of consuming resources and applications. Applications, resources and processing power are offered as a service by a provider and are available online via a Web browser. Software and data are also housed in data centers of providers. These are exploited in short and/or long periods depending on a payment model that depends on the effective consumption of resources. The processing power is paid based on the utilization of hardware resources (e.g. CPU) ; storage and data transfer are measured in terms of the amount of data stored or transferred. Over the last few years, the $\mathcal{CC}$ platform appeared as a new trend for managing data. Implicitly, resources in a Cloud can accommodate and/or process huge amounts of data. Most Cloud infrastructures are based on the distributed file system (DFS), whose main features are its ability for distributing and replicating data. The main files

B. Hong et al. (Eds.): DASFAA Workshops 2013, LNCS 7827, pp. 16–30, 2013.

and distributed storage systems of Cloud environment are GFS (Google File System) [1], HDFS (Hadoop distributed file system)[1], S3 (Simple Storage Service)[2] and Dynamo [2]. These systems are used at the data storage level. The execution level includes new models and paradigms dedicated to task execution. The most popular of these models is Google's *MapReduce* programming model [3] and its open source Implementation Hadoop.

With the growing reputation of $\mathcal{CC}$, many advanced applications such as decision supports are moving to the Cloud [4]. These applications are characterized by two dimensions: (i) they manage a large amount of data usually stored within a data warehouse ($\mathcal{DW}$) and (ii) they use complex queries to extract facts and knowledge in order to increase the decision making power. Due to the increased interest of small, medium and large companies towards $\mathcal{DW}$ technologies, commercial DBMS offered solutions, supports, tools and platforms for managing $\mathcal{DW}$. DFS, *MapReduce*, the extended traditional DBMS and NoSQL are examples of these solutions adapted to $\mathcal{CC}$. The *MapReduce* paradigm represents an advanced technology for processing a massive amount of data in a parallel way. This is due to its fault tolerance and its ability to operate in a heterogeneous environment. It may be seen as a complement to a DBMS for analytical applications [5]. Another aspect is its ability to implement relational operations.

To optimize complex decision support queries in the Cloud, optimization structures such as materialized views, indexes are required. Note that the selection of these structures in the classical environments such as centralized, parallel, distributed databases is based on cost models calculating the cost of executing these queries in the presence of the selected optimization structures. Usually, these cost models use logical parameters (size of database tables, attributes, selectivity factors of selection and join operations, etc.), physical parameters (the page size, the buffer size, etc.) and parameters of infrastructures (e.g. communication between nodes). Another parameter represents the query processing strategy such as join implementations (sort join, hash join, nested loop). In the Cloud environment, in addition to the traditional parameters related to the devices, other parameters related to the Cloud have to be included such as the map reduce paradigm of executing operations and the payment procedure.

Contrary to traditional databases, where large panoplies of cost models exist for each optimization structures, only few cost models exist for the Cloud. Their main limitation is that they do not consider all parameters. In this paper, we propose a cost model including query processing and payment parameters to select materialized views in the Cloud.

The rest of the paper is organized as follows: In Section 2, we give an overview of related work. In Section 3, we state a formalization of the materialized view selection problem in the cloud environment. In Section 4, our mathematical cost model is given, where all its components are detailed. Section 5 presents our experimental environment and the obtained results. Finally, in Section 6, we conclude the paper and outline issues for future works.

---

[1] http://hadoop.apache.org

[2] http://aws.amazon.com/s3

2   Related Work

Through the evolution of database technology, cost models played a crucial role. Initially they were used to estimate the cost of queries. A cost-based approach used by centralized/parallel traditional database optimizers (relational, object, etc.) to select query optimization plans is based on cost models. With the spectacular development of decision support systems, a large number of optimization structures are selected by the means of cost models [6–15]. However, the proposed analytical cost models, even those dealing with parallel $\mathcal{DW}$ [10, 11], are not suitable to be used in a $\mathcal{DW}$. This is because they do not take into account the features of the cloud environment, including payment model and data access methods used.

In the Cloud, to the best of our knowledge, only one cost model has been proposed recently by Nguyen et al. [16]. The authors have shown that the use of materialized views ($\mathcal{MV}$) in the Cloud reduces the cost of processing queries and they have proposed a theoretical cost model for data management in the Cloud, through the use of $\mathcal{MV}$ to optimize query performance. However, although this cost model treats the data storage in the Cloud and takes into account the payment model, it does not address the query evaluation manner as well as the physical features of nodes calculation, which have a significant impact on the cost of query processing.

In this work, we develop a new analytical cost model for designing a $\mathcal{DW}$ on the cloud in a *MapReduce* paradigm. This cost model allows us to estimate the execution cost of a query workload in the presence of materialized views.

3   Materialized View Selection Problem in the Cloud

DBMS have long been the standard technology for data warehousing, especially with the advent of parallel DBMSs which have a robust and highly efficient platform [17] based on the relational model. They also provide SQL as the standard language for expressing queries. The DFS and the *MapReduce* present a new technology for such applications [18]. They showed their performance to manage huge amount of data, where relational DBMS showed their limitations. This new massive parallel platform for data processing is considered the best candidate for complex data analysis (e.g. data mining) [5]. The *MapReduce* does not require a prior pattern of data and works well on unstructured and rarely changed data, whereas a DBMS is suitable for transactional queries and data sets that are structured, standardized and continuously modified. As a consequence, the *MapReduce* can be seen as a complement to a relational DBMS for analytical applications, since complex analysis problems require capabilities provided by both technologies [5].

Today, the $\mathcal{DW}$ offers several tools to analyze data: OLAP, data mining, statistics, etc. They are used in the field of data analysis, as they allow effective analysis of large amounts of data. Usually, a $\mathcal{DW}$ is typically modeled by a star schema [19] and requires complex and time consuming decision queries. Several optimization techniques are used to improve the performance of these queries, such as indexes and $\mathcal{MV}$. For each optimization object, several configurations are possible. The chosen configuration must be optimal or close to optimal.

The unlimited storage and the massive use of resources (storage, processing, etc.) performed on-demand are constrained to payment modes. During the selection of optimization structures in the Cloud, the storage constraint usually considered in the traditional databases is replaced by the payment cost that application owners have to pay, though expanding the storage space allocated to optimization structures can reduce the response time of decision queries. Therefore the payment cost to process queries may create a negative impact on the storage payment cost of optimization structures and then on overall cost.

Based on the above discussion, we define the problem of selecting optimization structures as follows: The problem of selecting optimization structures during the physical design phase consists in identifying a configuration that optimizes the execution cost of queries and/or the cost of total payment under a budget constraint. More precisely, we define the problem of selecting $\mathcal{MV}$ in the Cloud as follows: Let $V = \{v_1, v_2, ..., v_n\}$ be a set of candidate views, $Q = \{q_1, q_2, ..., q_m\}$ be the set of workload queries and B the budget for the processing of queries in the Cloud. The problem of selecting $\mathcal{MV}$ in the Cloud is to find a configuration $Config$ of $\mathcal{MV}$ such that: (i) the execution cost C in terms of query execution time and/or the treatment cost C' in terms of payment cost associated to the calculation of the workload queries is minimal, i.e.: $C_{Config(Q)} = Min(C_{V(Q)})$, $C'_{Config(Q)} = Min(C'_{V(Q)})$ and (ii) the sum of the payment cost of queries processing C_p in the presence of $\mathcal{MV}$, storage space cost C_s and maintenance cost C_m of all views $Config$ is less than the budget B, i.e: $\sum C_p(v_i) + C_s + \sum C_m(v_i) \leq B.$. In other words, the set of views minimizes the response time without violating the constraint B.

In *MapReduce* environments, pre-calculated queries are similar to materialized views in relational DBMS. *MapReduce* may be considered as a view definition and the calculated results as $\mathcal{MV}$ [20]. There are many similarities and differences between $\mathcal{MV}$ and *MapReduce* views (MRV), which are summarized in Table 1.

Table 1. Materialized views and pre-calculus *MapReduce* queries

	MV	MRV
	Calculated from one or more relations	Data extracted from one or more data sets
similarities	Stored in the $\mathcal{DW}$	Stored in DFS
	Defined by the user in SQL queries	Data transformation is defined by the user functions and Map Reduce
	Refresh views from data sources	Re-computation of *MapReduce* programs to obtain updates

4   Our Proposed Cost Model

In this section, we present our analytical cost model which consists in several mathematical formulas.

4.1   Estimation Assumptions and Parameters

Due to the complexity of the studied problem, we make assumptions on several aspects of the problem:

1. estimation of selectivity factors: we assume that data are uniformly distributed and attribute values are independent;
2. the CPU cost and the size of $tag - table$ are assumed to be negligible ;
3. all computation nodes are considered homogenous, i.e., they have the same physical features and their buffer size is large enough to hold the data.

Parameters of the payment cost are shown in Table 2. The physical features of nodes and the parameters and statistics on the $\mathcal{DW}$ (which is assumed to be modeled by a star schema) and query workload are shown in Table 3.

Table 2. Parameters for estimating the payment cost

Symbol	Description
P	Storage period
$NbrInst$	Number of instances used
$Price_{Inst}$	Price of one instance per unit of time (typically \$/h)
$Price_{Debit}$	Data transfer price (\$/GO)
$Price_{storage}$	Price of data storage per unit of time (\$/GO/U)

4.2   Processing Cost of the Query Workload

In a *MapReduce* system, such as Hive [21], a star join query between F and n dimensional tables runs in Nbr_{job} phases, where each phase corresponds to a *MapReduce* job [22]. The number of *MapReduce* jobs depends on the number of joins and the presence or absence of aggregation and sorting data. We can distinguish three cases: (1) the query contains only n successive join operations between the fact table F and n dimension tables (2) join operations are followed by aggregation and grouping operations on the results of the join (3) sorting is applied to the results. This number is given by the following formula:

$$Nbr_{job}(q) = n + x, \tag{1}$$

with $x = \begin{cases} 0 \text{ if the query contains only join operations ;} \\ 1 \text{ if the query contains aggregation and grouping operations ;} \\ 2 \text{ if the results are sorted.} \end{cases}$

Table 3. Parameters for estimating the query execution cost

Symbol	Description
M	Number of Map tasks
R	Number of Reduce tasks
D_i	Dimension table i
F	Fact table
$\|S\|$	Number of tuples in a relation S
$f(S,T)$	Join selectivity factor between the table T and S
N	Number of nodes used
D	Time required to transfer one byte
n	Number of dimension tables
m_i	Number of Map tasks that run on the node i
r_i	Number of Reduce tasks that run on the node i
Fr_q	Frequency of use of query q
$length(a)$	Length in bytes of attribute a
Sel_S	Selectivity of predicate defined on the relation S
$|a|$	Cardinality of attribute a
V	Set of $\mathcal{MV}$
K	Number of $\mathcal{MV}$
$TInit$	Average time to initialize a *MapReduce* job
$TFin$	Average time to close of a *MapReduce* job
$split$	Size in bytes of a split
$size(S)$	Size in bytes of a relation S
$Debit$	Data transfer rate (MB/s)
Rd	disk ratio
Rp	Ratio data replication
BS	Buffer Size
$Fupdate$	Frequency of update

Consequently, the total time of execution of a star join query is calculated by the following formula:

$$Texec(q) = \sum_{i=1}^{Nbr_{job}(q)} Tjob_i. \qquad (2)$$

Generally, a *MapReduce* job runs in two phases: Map and Reduce. Map outputs task are sent across the network to execution nodes of Reduce task (shuffle). To this end, we can calculate the execution time of a *MapReduce* job by the following formula :

$$Tjob_i = TInit + Tmap_i + Tshuffle_i + Treduce_i + TFin, \ i = 1, \ldots, Nbr_{job}(q) \quad (3)$$

The initialization time and the closing time of a *MapReduce* job are related to the initialization and closing of a job in the processing nodes. The initialization tasks of a job include for example receiving data and creating lists of tasks. The closure of a job is to establish a connection with the *JobTracker* in order to send the intermediate results, notifications of completion of the work, etc. Through experimentation, Pavlo et al. [17] found that *MapReduce* program takes some time before all nodes are functioning at full capacity. On a cluster of 100 nodes, it takes 10 seconds before the first Map task starts executing, and 25 seconds until all the cluster nodes start the execution.

We will now detail the cost of each term appearing in (3).

Map-Side Cost. We first give the general formula for the execution time of a Map phase, then we detail each term appearing in it. Since Map tasks are executed in parallel over all different nodes, the execution time of a phase Map is given by the following equation:

$$Tmap_i = \max_{j=1...N} m_j * \left(\frac{size(inMap_i)+size(outMap_i)}{M_i * Rd} \right), \quad i = 1,\ldots, Nbr_{job}(q),$$

where $M_i = \frac{inMap_i}{split}$, $i = 1,\ldots Nbr_{job}(q)$.

We recall that the size in byte of a relation S is given by the following equation: $size(S) = ||S|| * \sum length(a_j)$, where $\{a_j\}_j$ represents the set of attributes $\{a_1, a_2, ..., a_j\}$ that make up the relation S, and $||S||$ is the number of tuples of the relation S (following notations of Table 3).

The number of tuples $||inMap_i||$ is:

$||inMap_1|| = ||F|| + ||D_1||$ and $||inMap_i|| = ||RI_{i-1}|| + ||D_i||$, $i = 2,\ldots, n$, where RI represents intermediate results, RI_1 is the result of the first job, RI_2 the result of the second job and so on. If the query has aggregation and grouping operations, then: $||inMap_{n+1}|| = ||RI_n||$ and if the results are sorted, then $||inMap_{n+2}|| = ||RI_{n+1}||$.

The amount of output of one phase Map $||outMap_i||$ is given by the following formulas:

$||outMap_1|| = ||F|| * sel_F + ||D_1|| * sel_{D_1}$ and $||outMap_i|| = ||RI_{i-1}|| + ||D_i|| * \prod sel_{D_i}$, $i = 2,\ldots n$. If the query has aggregation and grouping operations, then $||outMap_{n+1}|| = ||RI_n||$ and if the results are sorted, then $||outMap_{n+2}|| = ||RI_{n+1}||$.

Finally, the number of tuples of intermediate results (RI) can be calculated using following formulas:

$||RI_1|| = ||F|| * sel_F * ||D_1|| * sel_{D_1} * f(F, D_1)$, and $||RI_i|| = ||RI_{i-1}|| * ||D_i|| * \prod sel_{D_i} * f(F, D_i)$, $i = 2,\ldots, n$, where we recall that $f(S, T)$ is the join selectivity factor between tables T and S. If the query has aggregation annd grouping operations, then

$||RI_{n+1}|| = Agreg(RI_n)$, with $Agreg(RI_n) = \prod |a_i| * sel_i$, where a_i represents a set of grouping attributes. If there is no Group By clause then $Agreg(RI_n) = 1$. If the results are sorted then:

$||RI_{n+2}|| = ||RI_{n+1}||$.

Note that The size in byte of a relation S may be given by the following equation:

$$size(S) = \sum_{j=1}^{||S||} length(a_j), \text{ where, } \{a_j\}_j \text{ represents the set of attributes of}$$
the relation S.

Reduce-Side Cost: We assume that Reduce tasks receive the same amount of data (balancing the workload between Reduce tasks). We can divide the cost of a Reduce stage into two costs : sorting cost (denoted C_1), which can be estimated by using the formula defined by Mishra et al. [23], and the cost of writing copies of the results in a DFS with replication rate Rp (denoted C_2). Therefore, the Reduce-side cost can be estimated by the following formulas:

$C1 = \frac{size(inReduce_i)}{R_i} * \log_{BS}\left(\frac{size(inReduce_i)}{R_i}\right)$, where BS is the buffer size, and $||inReduce_i|| = ||outMap_i||, \ i = 1, \ldots, Nbr_{job}(q)$.

Recalling that D is the time required to transfer one byte, we have

$C2 = D * Rp * \frac{size(outReduce_i)}{R_i * Rd}$, where $||outReduce_i|| = ||RI_i||, \ i = 1, \ldots,$ $Nbr_{job}(q)$. Finally, since tasks are executed in parallel over all nodes, we get

$$Treduce_i = \max_{j=1\ldots N} r_j * [C1 + C2], \tag{4}$$

$$\tag{5}$$

Shuffle Cost. We can estimate the time of data transfer between the Map tasks and Reduce tasks by the following equation :

$$Tshuffle_i = size(outMap_i) * D, \ i = 1, \ldots, Nbr_{job}(q). \tag{6}$$

Using the execution time given in formula (2) and the result transfer time, we are now in position to calculate the total cost of the query workload execution.

Total Time of the Query Workload Execution is equal to the sum of the execution time of all queries and transfer time multiplied by the query frequency Fr_q. Hence:

$Totaltime(Q) = \sum_{q \in Q}[(Texec(q) + Ttransfert(q)) * Fr_q]$, where $Ttransfert(q) = \frac{size(RI_{n+x})}{Debit}$.

4.3   Payment Cost of the Workload Queries Processing

Calculating the payment cost of a query q, $Cost_{finance}(q)$, is based on the execution time of the query , payment parameters shown in Table 2, and the cost of transferring the results of the query. This cost is given by the following formula:

$Cost_{finance}(q) = Texec(q) * NbrInst * price_{Inst} + Cost_{transfer}(q)$, where $Cost_{transfert}(q) = size(RI_{n+x}) * price_{Debit}$.

Total cost payment of the Query workload is given by the formula:

$TotalCost_{finance}(Q) = \sum_{q \in Q} Cost_{finance}(q) * Fr_q$, where Fr_q is the query frequency.

The execution cost of query workload in the presence of $\mathcal{MV}$ includes the following costs: (i) The storage cost of all $\mathcal{MV}$, (ii) the access cost to $\mathcal{MV}$ and (iii) the maintenance cost of all $\mathcal{MV}$. The storage cost of a view v_i is proportional to its size and storage period in the service provider and the price per unit of storage time (per hour, per day or per month). Therefore, $Cost_{storage}(v_i) =$

$size(v_i) * price_{storage} * P$, $i = 1, \ldots, K$., where K represents the number of materialized views.

The updates of the base relations may involve changes in the size of the $\mathcal{MV}$. Thus, the size of $\mathcal{MV}$ is not static. Suppose the size of v_i increases by an average ϕ_i each time unit. In this case, we can estimate the payment cost related to the storage cost for a period P by the following formula:

$Cost_{storage}(v_i) = \left((2 * size(v_i) + \phi_i * (P-1)) * \frac{P}{2} \right) * price_{storage}$, $i = 1, \ldots, K$,.

The cost of storing all $\mathcal{MV}$ is given by the following formula: $TotalCost_{storage}$ $(V) = \sum_{i=1}^{K} Cost_{storage}(v_i)$.

For computing the access cost of a query in the presence of $\mathcal{MV}$, two scenarios are considered:

Scenario 1: The execution of the query q implies the extraction of data for only a view v_i by a simple selection operation with aggregation, which corresponds to a single *MapReduce* job. Map phase filters data of v_i and Reduce phase calculates aggregations and returns the query results to HDFS. As before, we can calculate this cost by the following formulas:

$Texec(q, v) = TInit + Tmap_v + Treduce_v + Tshuffle_v + TFin + Ttransfert(q)..$

As before, we have $Tmap_v = \max_{i=1\ldots N} m_i \left(\frac{size(inMap_v) + size(outMap_v)}{M_v * Rd} \right)$, where $M_v = \frac{size(v)}{split}$, where $size(inMap_v) = size(v)$, $outMap_v = \|v\| * sel_v$. Recalling that D is the time required to transfer one byte, we have $Tshuffle_v = size(outMap_v) * D..$

The Reduce-side cost, $Treduce_v$, can be divided into two costs: the cost of sorting and cost of creating copies of the results in DFS. This cost can be estimated by using the formulas defined in section 4.2, so that:

$$size(inReduce_v) = size(outMap_v), \tag{7}$$

$$\|outReduce_v\| = Agreg(inReduce_v). \tag{8}$$

Scenario 2: The execution of a query q implies the join between a view v_i and p dimension tables, followed by aggregation and grouping operators. This requires $p + x$ MapReduce jobs, such that p first jobs calculate the join between v_i and p dimension tables and the latest jobs calculate the grouping, aggregation and sorting data. In this scenario, the cost of executing the query q can be given by the following formula:

$$Texec(q, v) = \sum_{j=1}^{p+x} Tjob_j, \quad v \in V. \tag{9}$$

In this case, $Tjob_j$, which is the execution time of a *MapReduce* job in the presence of $\mathcal{MV}$, is calculated by using the formula (3) by replacing the fact table F by the view v.

Thus, the Total access cost of query workload in the presence of $\mathcal{MV}$ can be estimated by the following formula:

$$Totaltime_{access}(Q, V) = \sum_{q \in Q, v \in V} [(Texec(q, v) + Ttransfert(q)) * Fr_q], \tag{10}$$

Furthermore, we can calculate the payment cost of processing queries in the presence of $\mathcal{MV}$. This cost depends on the processing time of the query and the type and the number of instances. This cost can be obtained by the following formula:

$$Cost_{finance}(q,v) = [Texec(q,v) * NbrInst * price_{Inst}] + cost_{transfert}(q), \ v \in V. \quad (11)$$

Therefore, the total payment cost of queries workload in the presence of the $\mathcal{MV}$ can be calculated by the following formula:

$$TotalCost_{finance}(Q,V) = \sum_{q \varepsilon Q} Cost_{finance}(q,v) * Fr_q, \ v \in V. \quad (12)$$

For the maintenance cost, we assume that the cost of maintaining a view v_i is equal to cost of re-computing it from its relevant sources. This maintenance is known as the static maintenance. The total cost of maintaining the set of materialized view V (denoted by $TotalCost(Q^M)$) is calculated as the sum of the re-execution cost of all queries Q'. This cost is estimated by the following formula:

$$TotalCost(Q^M) = \sum_{q_i \varepsilon Q'} [Texec(q_i) * Fupdate_{E_{q_i}}^{v_i}], \quad (13)$$

with $E_{q_i}^{v_i}$ is the set of basic relations involved by the query q_i computing the view v_i ($E_{q_i}^{v_i} \subset \{F, D_1, D_2, ..., D_n\}$) and $Texec(q_i)$ can be estimated by the function defined in the relation (2).

Re-execution of *MapReduce* programs of reconstruction the set V generates a cost of payment that can be estimated by the following formula:

$$TotalCost_{finance}(Q^P) = \sum_{q_i \in Q'} Cost_{finance}(q) \times Fupdate_{E_{q_i}}^{v_i}, \quad (14)$$

such that:

$$Cost_{finance}(q_i) = Texec(q_i) * NbrInst * price_{Inst}, \ q_i \in Q'. \quad (15)$$

5 Theoretical and Practical Results

This section presents an experimental study in a *MapReduce* system through which we deploy a $\mathcal{DW}$ issued from the star schema benchmark (SSBM) [24] on Hive and we execute queries on its data.

5.1 Assimilation of Infrastructure

In order to use a business intelligence solution based on the concept of the cloud, we have assimilated the infrastructure described in Figure 1, where the hardware architecture consists of commodity machines associated with virtual machines with GNU/Linux operating system. In our work, we have used an Ubuntu virtual machine equipped with the operating system Ubuntu-11.10 with Linux kernel version 2.6.27-11-server and with a processor of 2.10 GHz and memory 1 GB, with a *MapReduce* system: Hive, HDFS, and Hadoop.

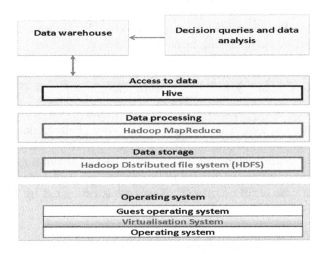

Fig. 1. Infrastructure of the BI Cloud Computing solution

5.2   Data Warehouse Benchmark

The schema of our benchmark is composed of one fact table ($LINEORDER$) and four dimension tables ($CUSTOMER$, $SUPPLIER$, $PART$, and $DATE$). Depending on the capacity of the used system, we have generated data with $SF = 1$, which gives us a total size $\simeq 570$ MB of data. In our experiments, we consider three queries of the benchmark: $Q1.1$, $Q3.1$ and $Q4.3$. Small adjustments to fit with the assumptions defined in section 4.1 are added.

5.3   Obtained Results

In this section, we detail our experimental methodology. As we said before, our experiments are performed using our theoretical cost model and on the Hive.

In the Hive environment, first of all, these three queries are transformed into HiveQL and then executed on the Hive. The obtained queries are described as follows:

```
CREATE TABLE Q1.1(revenu BIGINT) INSERT OVERWRITE TABLE Q1
SUM(1.lo extendedprice * 1.lo discount) AS revenue
FROM Lineorder l JOIN Dat d
ON l.lo orderdate = d.d datekey and d.d year=1993 and l.lo discount >= 1 and
l.lo discount<= 3 and l.lo quantity < 25 ;
```

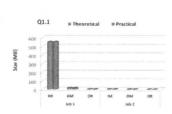

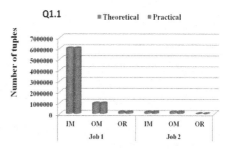

Fig. 2. The size of intermediate results of Q1.1. jobs

Fig. 3. The number of IO of jobs related Q1.1

```
CREATE TABLE Q3.1(nation1 STRING, nation2 STRING, year INT, revenue BIGINT)
INSERT OVERWRITE TABLE Q3
SELECT c nation as nation1, s nation AS nation2, d year AS year, SUM(lo revenue) AS
revenue
FROM Lineorder l JOIN Dat d
ON l.lo orderdate = d.d datekey and d.d year>=1992 and d.d year<=1997
JOIN supplier s
ON l.lo suppkey=s.s suppkey and s.s region='ASIA'
JOIN Customer c
ON l.lo custkey=c.c custkey and c.c region='ASIA'
GROUP BY c nation, s nation, d year
ORDER BY year ASC, revenue DESC;

CREATE TABLE Q4(year int, profit BIGINT, city STRING, brand STRING)
INSERT OVERWRITE TABLE Q4
SELECT d.d year as year, SUM(l.lo revenue-l.lo supplycost) AS profit, s.s city AS city,
p.p brand AS brand
FROM lineorder l JOIN Customer c
ON l.lo custkey = c.c custkey and c.c region='AMERICA'
JOIN Supplier s
ON l.lo suppkey = s.s suppkey and s.s nation='UNITED STATES'
JOIN Part p
ON l.lo partkey=p.p partkey and p.p category='MFGR#14'
JOIN Dat d
ON l.lo orderdate = d.d datekey and d.d year<> 1992 and d.d year<> 1993 and d.d year<>
1994 and d.d year<> 1995 and d.d year<> 1996
GROUP BY d.d year, s.s city, p.p brand
ORDER BY year, city, brand
```

Materialized views are selected using Yang et al. algorithm [15]. This algorithm exploits the interaction between queries. Afterwards, the queries are executed on Hive with the presence of these materialized views and their execution costs are computed.

For the theoretical point of view, the cost of the same queries is estimated by the use of our cost models.

For each query job (corresponding either to a join or an aggregation), we compute the size of intermediate results (in MB) and the number of IOs (in terms of tuples) for the Map and Reduce phases.

Figures 2, 3, 4, 5, 6, 7 compare the results obtained by the use of our theoretical cost model and our platform. Note that IM, OM, IR and OR represent respectively *inMap, outMap, inReduce and outReduce* costs. We notice that our cost model gives a good approximation of the I/O cost. Estimated cost values are slightly under-predicted or over-predicted compared to the actual values. This can be attributed to several reasons.

We assumed as assumption in section 4.1 that the data are uniformly distributed. However, in practice this is not always the case. Thus, during the

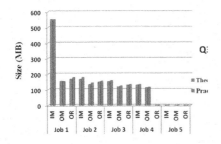

Fig. 4. The size of intermediate results of Q3.1. jobs

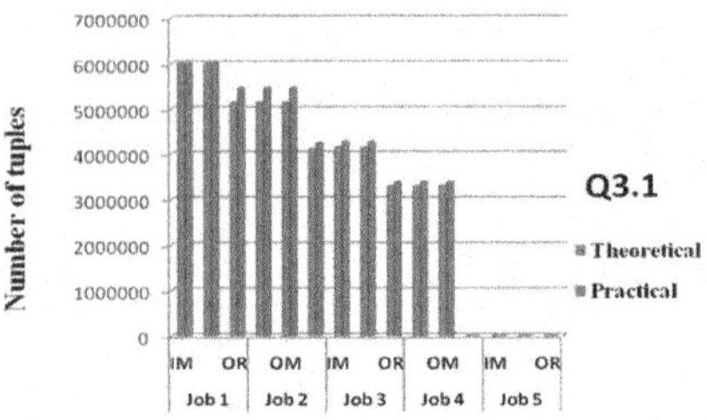

Fig. 5. The number of IO of jobs related Q3.1

creation of intermediate files, the framework adds control data that increase the amount of data read or written. Finally, the experimental results indicate that our theoretical cost model is able to evaluate the execution plan of decision queries in terms of I/O cost in a *MapReduce* environment with a good approximation.

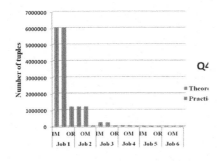

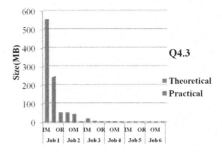

Fig. 6. The size of intermediate results of Q4.3 jobs

Fig. 7. The number of IO of jobs related to Q4.3

6   Conclusion

In this work, we have presented a *preliminary work including a new analytical cost model for physical design* of a $\mathcal{DW}$ in the Cloud. This model includes a *large set of mathematical formulas* that take as input parameters, statistics on the data and queries workload and features of the Cloud, and return the results as cost generated by the query processing decision on the Cloud. The model developed is characterized by its simplicity and ease of implementation and allows taking into account the features of Cloud, particularly the payment model and execution environment. Thus, it is able to conduct the algorithms of selection of the $\mathcal{MV}$ on the Cloud in a *MapReduce* environment and compare between the materialization of the query and the increase in number of compute nodes.

Furthermore, it can be used to choose the best execution plan for a query in such environment.

As a perspective, we are currently studying the scalability of our solution. Note that this work is fruit of the final year project of the master degree of Assia Brighen, done in Bejaia University -Algeria.

References

1. Ghemawat, S., Gobioff, H., Leung, S.T.: The google file system. In: Proceedings of the 19th ACM Symposium on Operating Systems Principles (SOSP), pp. 29–43 (2003)
2. DeCandia, G., Hastorun, D., Jampani, M., Kakulapati, G., Lakshman, A., Pilchin, A., Sivasubramanian, S., Vosshall, P., Vogels, W.: Dynamo: amazon's highly available key-value store. In: Proceedings of Twenty-First ACM SIGOPS Symposium on Operating Systems Principles (SOSP), pp. 205–220 (2007)
3. Dean, J., Ghemawat, S.: Mapreduce: simplified data processing on large clusters. Communications of the ACM 51, 107–113 (2008)
4. Abadi, D.J.: Data management in the cloud: Limitations and opportunities. IEEE Data Engineering Bulletin 32(1), 3–12 (2009)
5. Stonebraker, M., Abadi, D.J., DeWitt, D.J., Madden, S., Paulson, E., Pavlo, A., Rasin, A.: Mapreduce and parallel dbmss: friends or foes? Communications of the ACM 53(1), 64–71 (2010)
6. Aouiche, K., Darmont, J.: Data mining-based materialized view and index selection in data warehouses. Journal of Intelligent Information Systems 33(1), 65–93 (2009)
7. Aouiche, K., Jouve, P.-E., Darmont, J.: Clustering-based materialized view selection in data warehouses. In: Manolopoulos, Y., Pokorný, J., Sellis, T.K. (eds.) ADBIS 2006. LNCS, vol. 4152, pp. 81–95. Springer, Heidelberg (2006)
8. Baril, X., Bellahsene, Z.: Selection of materialized views: A cost-based approach. In: Eder, J., Missikoff, M. (eds.) CAiSE 2003. LNCS, vol. 2681, pp. 665–680. Springer, Heidelberg (2003)
9. Bauer, A., Lehner, W.: On solving the view selection problem in distributed data warehouse architectures. In: Proceedings of the 15th International Conference on Scientific and Statistical Database Management (SSDBM), pp. 430–451 (2003)
10. Bellatreche, L., Benkrid, S.: A joint design approach of partitioning and allocation in parallel data warehouses. In: Pedersen, T.B., Mohania, M.K., Tjoa, A.M. (eds.) DaWaK 2009. LNCS, vol. 5691, pp. 99–110. Springer, Heidelberg (2009)
11. Bellatreche, L., Cuzzocrea, A., Benkrid, S.: $\mathcal{F}\&\mathcal{A}$: A methodology for effectively and efficiently designing parallel relational data warehouses on heterogenous database clusters. In: Bach Pedersen, T., Mohania, M.K., Tjoa, A.M. (eds.) DAWAK 2010. LNCS, vol. 6263, pp. 89–104. Springer, Heidelberg (2010)
12. Bellatreche, L., Boukhalfa, K., Richard, P., Woameno, K.Y.: Referential horizontal partitioning selection problem in data warehouses: Hardness study and selection algorithms. International Journal of Data Warehousing and Mining (IJDWM) 5(4), 1–23 (2009)
13. Chan, G., Li, Q., Feng, L.: Optimized design of materialized views in a real-life data warehousing environment. International Journal of Information Technology 7(1), 30–54 (2001)
14. Golfarelli, M., Rizzi, S.: Methodological framework for data warehouse design. In: ACM First International Workshop on Data Warehousing and OLAP (DOLAP), pp. 3–9 (1998)

15. Yang, J., Karlapalem, K., Li, Q.: Algorithms for materialized view design in data warehousing environment. In: Proceedings of 23rd International Conference on Very Large Data Bases (VLDB), pp. 136–145 (1997)
16. Nguyen, T.V.A., Bimonte, S., d'Orazio, L., Darmont, J.: Cost models for view materialization in the cloud. In: EDBT/ICDT Workshops, pp. 47–54 (2012)
17. Pavlo, A., Paulson, E., Rasin, A., Abadi, D.J., DeWitt, D.J., Madden, S., Stone-braker, M.: A comparison of approaches to large-scale data analysis. In: SIGMOD Conference, pp. 165–178 (2009)
18. Ordonez, C., Song, I.Y., Garcia-Alvarado, C.: Relational versus non-relational database systems for data warehousing. In: ACM 13th International Workshop on Data Warehousing and OLAP (DOLAP), pp. 67–68 (2010)
19. Inmon, W.H.: Building the Data Warehouse. John Wiley & Sons, Inc, New York (2005)
20. Jörg, T., Parvizi, R., Yong, H., Dessloch, S.: Incremental recomputations in mapre-duce. In: Proceedings of the Third International Workshop on Cloud Data Management (CloudDB 2011), pp. 7–14 (2011)
21. Thusoo, A., Sarma, J.S., Jain, N., Shao, Z., Chakka, P., Anthony, S., Liu, H., Wyck-off, P., Murthy, R.: Hive - a warehousing solution over a map-reduce framework. PVLDB 2(2), 1626–1629 (2009)
22. Han, H., Jung, H., Eom, H., Yeom, H.Y.: Scatter-gather-merge: An efficient starjoin query processing algorithm for data-parallel frameworks. Cluster Computing 14(2), 183–197 (2011)
23. Mishra, P., Eich, M.H.: Join processing in relational databases. ACM Computing Surveys 24(1), 63–113 (1992)
24. O'Neil, P., O'Neil, B., Chen, X.: Star schema benchmark (2009)

BC-BSP: A BSP-Based Parallel Iterative Processing System for Big Data on Cloud Architecture

Yubin Bao[1], Zhigang Wang[1], Yu Gu[1], Ge Yu[1], Fangling Leng[1],
Hongxu Zhang[2], Bairen Chen[2], Chao Deng[3], and Leitao Guo[3]

[1] School of Information Science and Engineering, Northeastern University, Shenyang, China
{baoyubin,guyu,yuge,lengfangling}@ise.neu.edu.cn
[2] Software Division, Neusoft Corp., Shenyang, China
{kevinzhang,chenbr}@neusoft.com
[3] China Mobile Institute, China Mobile Corp., Beijing, China
{dengchao,guolt}@chinamobile.com

Abstract. Many applications in real life can produce and collect large amount of data and many of them can be modeled by Graph. The number of vertexes of a graph could be several hundreds of millions to billions and the number of edges could be ten or more times of the number of its vertexes. A BSP-based system for large-scale data (especially graph data) parallel and iterative processing is discussed in this paper. The system has the ability to flexible configuration and the extendibility for functions and strategies (such as adjusting the parameters according to the volume of data and supporting multiple aggregation functions at the same time), to process large-scale data, to tolerate faults, to balance load, and to run clustering or classification algorithms on metric datasets. Lots of experiments are done to evaluate the extendibility of the system implemented in the paper, and the comparison between BC-BSP-based applications and *MapReduce*-based ones are made. The experimental results show that BSP-based applications have higher efficiency than that of *MapReduce*-based applications when the volume of data can be put in the memory during the course of processing; on the contrary the latter are better than the former, and the performance of BC-BSP platform outperforms *Hama* and *Giraph*.

Keywords: BSP, MapReduce, Graph Processing, Disk Cache, Big data.

1   Introduction

Graph is an abstract data structure which has been researched deeply in the area of computer science. It is so common to express the real world using graph, such as the road network, the spread of disease, the reference among technological literature, the links among web pages, the relationship among all kinds of objects in social network and the biological information network. So graph model can be used widely to model many applications. In spite of the theory and algorithms on graph have been researched in depth during the past several decades, most of them focus on small-scale datasets. With the development of the information technology, the scale of all kinds of

B. Hong et al. (Eds.): DASFAA Workshops 2013, LNCS 7827, pp. 31–45, 2013.
© Springer-Verlag Berlin Heidelberg 2013

information keeps increasing rapidly, which leads to the scale of graphs larger and larger. The number may be even high in social network. Such as *Facebook*, the largest scale social network has about 700 million users. For search engines, such as *Google* and *Baidu*, it is necessary to evaluate web pages importance by related algorithms. The most famous one is *PageRank* algorithm. We can define a web page as a node in a link graph and the link between two pages is regarded as an edge with direction. So the rank score of a web page can be computed according to the links among pages. Given that the graph is organized by adjacent list and one whole record needs 100 bytes to store, if we store 10 billion nodes and 60 billion edges, the whole storage space will be more than 1 TB. The situation is similar with other applications, such as social network. The cost of time and space during processing the big scale graph has already exceeded the ability of concentrated computing traditionally. In conclusion, it has become a new challenge to process large scale data, especially large scale graph efficiently.

At present, *MapReduce*[1] computing model based on *Hadoop* ecosystem can process large-scale graph data with better fault-tolerance and scalability. While, most graph algorithms need to process graph data many times iteratively. One or more jobs are needed to complete an iterative computing task. As we known, the cost of the warm-up start of a *MapReduce* job is considerable. In order to solve this problem, *Google* developed an system for large-scale graph processing based on BSP model, called *Pregel*[2]. *Pregel* can process graph data in parallel and implement the communication among workers by message passing. However, *Pregel* assumes that all data including the processed data and intermediate data (such as message data) is resident in memory during the processing. Apparently, if the number of workers and the main memory capacity of each worker machine are limited, the scalability is also limited. It is not an open source project. There are two open source projects based on BSP model, *Hama*[3] and *Giraph*[4]. *Hama* is also good at processing big data iteratively, especially for processing matrix. But it does not consider the disk as an assistant device to temporary store graph data or messages when main memory is overflowed too in its early version. *Giraph* developed by *Yahoo* implements the BSP model based on *Hadoop* framework. Simply speaking, an application on *Giraph* is a special *MapReduce* job without reduce stage. It designs an inbuilt loop in the map task to simulate the super-steps of BSP model.

In this paper, we design a system BC-BSP in the Big Cloud environment of China Mobile Corp. Therefore we call it BC-BSP (Big Cloud-BSP), which is good at iteratively processing large scale graph data and other structured data. The features of BC-BSP and our contributions are as follows. a) BC-BSP implements the BSP model and uses the disk as the swap space to store part raw data and some intermediate data during the iterative processing when they all can not be put in main memory. So we can handle relative larger scale graph data if the number of available workers and the total memory capacity are limited. b) It provides flexible configuration and scalability. Users can choose or define the format of input and output. BC-BSP supports many data format, such as distributed file system, database. If it is need, users can define the special data format by relative interface. BC-BSP also supports a lot of strategies to partition the raw data. BC-BSP supplies hash partition, local partition and user-defined partition. c) It takes load-balance into consideration. BC-BSP schedules tasks to workers with the consideration of data locality and tries to keep the load-balance. Especially, the load balance among workers is more prior than the data locality. d) Some experiments are made to compare and evaluate the performance and scalability between the applications based on BC-BSP and *MapReduce*.

The rest parts of the paper are arranged as follows. Section 2 introduces BSP model, section 3 gives the overview of BC-BSP, section 4 describes the interfaces of BC-BSP, the implementation of BC-BSP is presented in section 5, section 6 presents two application examples, *PageRank*[5] and *K-means*, on BC-BSP, section 7 shows the experimental results and the analysis about them, and the last section draws the conclusions and discusses some points and the future work.

2   Introduction to BSP Model

BSP[6](Bulk Synchronous Parallel) is a "bulk" synchronous model. There is a master to coordinate the whole other workers, which are the nodes in the cluster for storing data and running program to process data. BSP model is a parallel computing model based on super-step. A BSP-based application consists of a series of super-steps (see in Fig. 1(a)). In each super-step, the tasks on the cluster workers are asynchronous parallel running, and they can send messages to other workers for satisfying the requirements of the computing job. The next super-step can start until the computing of each worker has ended and messages sending and receiving of each task has completed. It is called barrier synchronization (see in Fig. 1(b)).

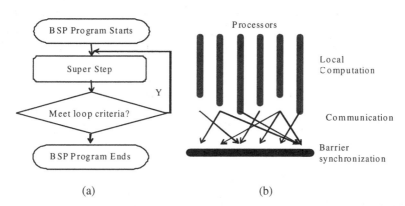

(a)   (b)

Fig. 1. BSP model. (a) The macro-running procedure of a BSP program, (b) The running procedure in a BSP super-step

3   Overview of BC-BSP

Figure 2 shows the framework of BC-BSP system which consists of the core computing engine and management tools. The core computing engine consists of the *Client*, the *BSPController*, the *Worker*, the *Task*, the *Global Synchronizer*, the *Message*

Communicator, API/CLI, and the *Fault Tolerance Controller*. The management tools consist of deployment and configuration tool, log management tool, performance management tool, and fault management tool.

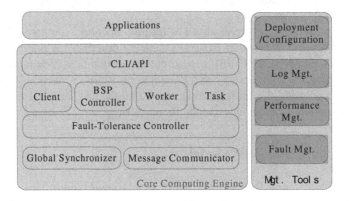

Fig. 2. The entire system framework of BC-BSP

The *Client* splits the input data according to the input path given by users, adjusts the number of partitions which is the processing unit of each task in a worker, asks the *BSPController* for the job ID, packs the job, and then submits the package to the *BSPController*. After the job begins, it is also responsible for reporting the running status in time. The *BSPController* manages the registration of the worker nodes in the computing cluster, the heartbeats from each worker, the status information of the cluster, and acts as a control centre of the fault-tolerance control. It also provides all the interfaces for the status query. It is responsible for scheduling, initialization, running monitor, and synchronization control of the jobs. The *Worker* manages the local jobs, local synchronization control, and local aggregation on a worker node. The *Task* is the entity that runs the jobs, and is responsible for inputting and outputting data and processing the local data. The *Global Synchronizer* manages the global synchronization among all the workers in each super-step by using *Zookeeper*, which is an open source middleware for a centralized service for providing distributed synchronization and et al. The global synchronization of a super-step is completed by the *BSPController*, the workers, and the tasks in cooperation. During the synchronization, the aggregation can be completed by invoking the aggregation function provided by users. The *Message Communicator* is responsible for sending and receiving messages, and for caching the messages received from other tasks to the local queue of received messages which can be saved into disks temporally when the main memory is not enough during the processing in every super-step. The *Fault-Tolerance Controller* detects faults, backups the snapshots for fault-tolerance, and recovers the system from failures. It uses the checkpoint mechanism for fault-tolerance. The *CLI/API* provides application program interfaces for local computation, sending or receiving messages, and etc. It also provides the commandline interface for the startup and shutdown of the system service, submitting the jobs, and manually specifying checkpoint and etc. The *Management Tools* uses the web interface or visual interface to provide users a method to manage the system.

The *Deployment/Configuration* tool provides users a visual interface for deploying system and configuring each node in the computing cluster. *Log (Fault) management* tool is used to check running logs (faults occurred during the running of a job) of the running jobs and the completed jobs. *Performance management* tool is used to monitor the system running status and the job running status.

Figure 3 shows the control mechanism of the BC-BSP's running. It shows the collaborative relationship among the *Client*, the *BSPController*, the *Workders*, the *Tasks* and the *Zookeeper*. Users interact with the BC-BSP system by *Client*, such as submitting jobs and monitoring the job running status. *BSPController* is the central nervous system of the entire BC-BSP system, and is responsible for controlling the whole cluster. The *WorkerManager* is the control center of a worker node, and manages the running and controlling of the worker node including collecting the information of all the tasks of a job on the worker node, communicating with the *BSPController* and with other workers. The *Task* is the work entity which performs the specific computing work. A job may have several tasks running on one worker, and these tasks are managed by the *WorkerAgentForJob* on this worker. The synchronization during the system running is controlled by the *Zookeeper*. One worker runs one *WorkerManager* process while there are several jobs running on it. Therefore, at the same time one *WorkerManager* may consist of several *WorkerAgentForJob* objects which manage all the tasks of a job on this worker.

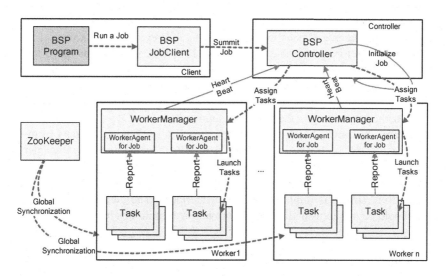

Fig. 3. The internal structure of BC-BSP

4   The Interfaces

BC-BSP system provides two types of interface, command-line interface and APIs (Application Program Interface). We mainly introduce the APIs.

When users write their application programs, they can invoke the APIs to extend BSP functions to meet their needs. For example, they can use *Combiners* interface to specify a combiner function for merging messages, use *Aggregators* interface to perform aggregation function, use *Partitioner* interface to control the partitioning of the input data, use *ContextInterface* interface to pass messages and execute local computation. The following is a brief introduction to these APIs.

VertexContextInterface: It is used to supply the context of a vertex for message passing and computing. At each super-step in a certain job, the system needs the related attributes of the vertex which is being processed, including vertex ID, value and current aggregation results, the update of vertex value and each edge value, the incoming and outgoing messages. So, these attributes and the operation methods on these attributes are encapsulated in this interface.

Combiners: During the graph processing, graph vertex is processed one by one. At each super-step, a vertex sends messages to its adjacent vertices and receives messages which are sent by other vertices at the same time. *Combiners* are used to merge messages at the sender side to reduce communication overhead. And different applications may require different combine function, for example, *sum*(), or *count*(). Therefore, users can specify their own combiner function to merge messages by extending this interface.

Aggregators: The graph processing needs aggregation in many cases, e.g., in order to examine whether the iteration should be stopped, *PageRank* application needs to aggregate the rank errors between the current super-step and last one. So, users can specify their own aggregators by implementing the *Aggregators* interface.

Partitioner: Before processing the data, the raw data should be assigned to each task by a certain principle. The default *Partitioner* provided by the system is hash function based on *hashCode*() method of Java. The *getPartitionID*() method in *Partitioner* interface maps a vertex ID into the corresponding partition ID. Users can override that method according to their own requirements.

Input and Output: The Input Interface is used to read the graph data from data source, e.g., *HDFS* or *HBase*. Therefore, *RecordReader* and *InputFormat* interface are provided for defining the input format like *Hadoop* and users can implement them to specify an input format to meet their needs. For example, the input format used to read data from *HBase* is implemented these interfaces. The output of the processed results should be output using the output format. Its definition is like to the input.

5   Implementation of BC-BSP

This section will introduce some implementation strategies and details of BC-BSP system. That includes the format of graph data, *BSPController* which is in the *Controller* node, *WorkerManager* which is the manager of a worker, *Task* which executes the *compute*() provided by user, passes messages, controls global synchronization and fault tolerance under the computing framework of BS-BSP system.

5.1   The Presentation of Graph

The system is mainly for processing large scale graph data, but it can also process structured data with the same data type elements. So, we mainly introduce the presentation structure of graph data. The graph is made up of vertex collection and edge collection, so there are *Vertex* class and *Edge* class to present the graph data. BC-BSP adopts the adjacent list to organize the graph data. In the *Vertex* class, there are some vertex attributes (such as vertex ID) and the information on outgoing edges. Meanwhile, it supports related methods to operate the member variables (see Fig. 4).

```
public class Vertex implements Writable {
    String vertexId = null; // vertex ID
    String vertexValue = null; // vertex value
    List<Edge> outEdgeList = new ArrayList<Edge>(); //store outgoing edge information
    public void addOutEdge(Edge outEdge) { }; // add an outgoing edge
    public List<Edge> getAllOutEdge() { }; // get all outgoing edges
    public void setAllOutEdge(List<Edge> aoutEdgeList) { }; // set all outgoing edges
    public boolean removeEdge(Edge edgeNode) { }; // remove an outgoing edge
    public boolean updateEdge(Edge edgeNode) { }; //update an edge value of the vertex
    public int hashCode() { }; // computing the hash code of a vertex
    public int getOutEdgeNumber() { }; // get # of outgoing edges from the vertex
    ......
}
```

Fig. 4. The structure of graph vertex class

The above structure of graph vertex class is suitable for graph data. However, it also can describe structured data by converting them to fit the above structure. For example, we can treat a record of the raw structured data as the vertex value string. Therefore, each record of raw structured data as a graph vertex.

5.2   BSPController Implementation

The *Controller* node is the center of the whole BC-BSP cluster. From the hardware perspective, it is responsible for managing all the worker nodes; from the software point of view, it is responsible for monitoring the working status of the computing cluster, receiving heartbeat information from each worker and process it, controlling the global synchronization among workers for each job. When the cluster starts, the *Controller* node receives registration information from each node to form unified cluster resource information. During the course of normally working, *Controller* collects and updated the cluster resource information (such as the number of free task slots) by the heartbeat mechanism periodically. When a user requests to submit jobs, the *Controller* assigns a unique job ID to the job, and then generates a job control object and puts it into the job waiting queue. The job scheduler selects high priority jobs to run in accordance with the priority and FIFO strategy. Then the task scheduler assigns tasks to workers in accordance with the principle of load balance and data locality. Because all the tasks for a job need to run at the same time, the task scheduler pushes down the tasks to each worker node, not like the scheduling way in *Hadoop* ecosystem, in which worker node applies a task to run when it has free task slots.

5.3   Worker Manager

A worker node is a unit of computation. After starting of BC-BSP, each worker node of the BC-BSP cluster will create a *WorkerManager* process which manages the tasks and maintains the synchronization among them. As soon as the startup of each *WorkerManager*, it should register itself to the *BSPController* to join the BC-BSP computing cluster. During their life time, they send heart-beat signals to the *BSPController* to report their status, respectively. When a new task arrives at a worker, *WorkerManager* reads and unpacks job profiles from *HDFS* into the local file system, and creates a *TaskinProgress* object and task process, and finally starts the task process. *WorkerManager* will build a *WorkerAgent* object for the tasks, which are running on the same node, to collect the status information of the task for a job. In this case, two levels of synchronizations are needed. The low level synchronization is for all tasks belong to the same job on the same worker node to synchronize, and then the high level is to register to *ZooKeeper* by worker as a whole for synchronization. So, the two synchronization levels decline the amount of client-ends keeping connection with *ZooKeeper* server and then decrease the workload of *ZooKeeper*. *WorkerManager* manages the tasks belong to the same job as a whole, so that it can manipulate some local computing, like the aggregation of local results computed by local tasks.

5.4   Task and Message Passing with Disk Assistant

A task is a logic computing unit. Task scheduler in *BSPController* assigns tasks to worker nodes according to load balance and data locality, and then *WorkerManager* on worker node creates task processes. After task starting, it first loads data processed by it, that is, it reads data from storage media according to the specified input format and then partitions the data into different partitions. During the course of partitioning, some graph vertex data need to be transferred to other tasks. After the completion of data partitioning, a global synchronization is needed to wait for all tasks belonging to a job to finish data partitioning. Then, tasks can go into BSP's super-steps to process graph data iteratively, that is, local computing, message passing and global synchronization. During the computing, task may send heart-beat information periodically to *WorkerAgent* object in *WorkerManager* process to report its current status.

Pregel system and its different implements all suppose that there are enough worker nodes and resources in the cluster to hold all graph data processed in a task and the related intermediate data (such as messages) during the course of each super-step in main memory completely. But actually, this assumption doesn't hold. There are two-folder reasons. The first folder is that it is difficult for a user to determine how many workers to be used and whether there are enough main memory to hold the graph data and the messages for a given dataset. The second is that the system can handle relative large-scale data under the limitation of the cluster scale.

For the above reasons, BC-BSP system applies disk space to store some graph data and intermediate messages temporally in order to process relative large-scale data. BC-BSP divides the JVM heap space into three parts. They are the spaces used by temporary defined objects, the spaces for storing graph data objects, and the spaces for messages. The space percentages occupied by the three parts are α, β and γ, respectively. The sum of these parameters is equal to 1. So, user only needs to give any two ones, and their values can be given in configuration file according to the real situation of the processed data.

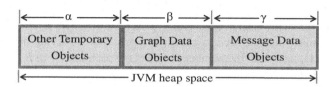

Fig. 5. The management model for JVM heap space

In order to avoid the overflow of memory, the data should be spilled into disk when the memory occupied by the data exceeds the given threshold. In our system, no matter graph data or message data both are swapped applying hash bucket techniques. In this case, the hash bucket for graph data objects and the one for message data objects have one to one relationship. Therefore, the super-step computing of a task can process graph data objects in hash bucket one by one, at the same time the message objects are all in a hash bucket related to the hash bucket holding graph data objects. Therefore, the system can quickly match the graph data object with the messages sent to it.

For message objects, each task maintains three queues. The one is *IncomedQueue* for managing the messages which are sent from the last super-step, processed in the current super-step, and is kept in memory as far as possible. The second one is *IncomingQueue* for managing the messages which are sent from other vertexes in the current super-step, will be processed in the next super-step, and has the highest priority to be spilled into disk. The third one is *OutgoingQueue* for managing the messages which are produced during the computing, is never spilled into disk. Messages in this queue can be combined by invoking the *combine()* method defined by user when the length of the queue exceeds a given threshold, and will be sent to other tasks. Practically, these message queues can be organized by hashmap.

5.5 Data Partitioning

Each task calls data partitioning function to read the binding data split from a specified data source, and uses the data partitioning strategies, such as hash partitioning, to allocate the graph data to a partition which is processed by a task. The system provides *hashCode()* method of Java as a default hash function, and also provides an interface for users to define their own hash functions to meet their special partitioning requirements. The input of the hash function is the value of a vertex ID, which can be an integer or a string; the output of the hash function is the partition ID. Therefore, a map table between PartitionID and Worker can be established to record a partition on which worker node. We can use MD5 method, which is a widely used hash algorithm for getting digest information of the input string, to partition graph data in order to get balanced distribution. But in practice, we find that its time consuming is very large. There are two reasons. The first is the MD5 computing of an input string needs more time than that of *hashCode*() method, the second is that the MD5 value should be computed for each graph vertex frequently during each super-step.

Whether the size of each data partition is equalized approximately will make a direct impact on the system load balance and the performance. It is known that hash map is difficult to ensure the equilibrium of each partition. To this end, we use the division method of multi-hash buckets merge to achieve the load balance. The basic idea is that assuming that we need to get *n* partitions, first we divide input data into *k**n buckets ($k \geq 1$), then send the number of objects in each bucket to *BSPController*, who merges *k**n buckets into *n* buckets according to load-balance principles. The merge principles can make the data objects in each bucket as possible as balance; it can also consider the data locality.

6   Application Examples

We design and implement several applications using the *APIs* supplied by the system, such as *PageRank*, *SSSP*(Single Source Shortest Path), and *K-means* on non-Graph structured data. But only the *PageRank* algorithm based on BSP is discussed on detail and K-means algorithm is discussed briefly because of the space limitation.

6.1   PageRank

Fig. 6 describes the implementation of *PageRank* algorithm based on BS-BSP platform. *PageRank* algorithm needs send the current rank value of the vertex to its adjacent vertexes which are linked by the current vertex by some rules (such as equal allocation) as the contribution value to the adjacent vertex. According to the *PageRank* algorithm, the messages send to the same vertex can be merged by the way of summary. Therefore, we design and implement a *combine*() method by overriding the *combine*() method of *Combiner* interface class (see in Fig. 7).

```
import com.chinamobile.bcbsp.*;
public class PageRankBSP extends BSP {
    ...... // omitted some variable definition
    @Override
    public void compute(Iterator<BSPMessage> messages, BSPStaffContextInterface
                    context) throws Exception {
        /* Receive messages sent to this vertex */
        receivedMsgValue = 0.0;
```

```
receivedMsgSum = 0.0;
while (messages.hasNext()) {
    receivedMsgSum+=Double.parseDouble(new
                        String(messages.next().getData())); }
/* Process received messages and Update vertex value */
if (context.getCurrentSuper - stepCounter() == 0) {
  sendMsgValue = Double.valueOf(context.getVertexValue())
                        /context.getOutgoingEdgesNum();
} else {
      /* According to the sum of error to judge the convergence */
      errorValue=(ErrorAggValue)context.getAggregateValue(ERROR_SUM);
      if (Double.parseDouble(errorValue.getValue())< ERROR_THRSHLD) {
          context.voltToHalt();   // This vertex can halt
          return; }
    /*Compute new vertex rank value and the contribution to adjacent vertex*/
    newVertexValue=CLICK_RP*FACTOR+receivedMsgSum*(1- ACTOR);
    sendMsgValue = newVertexValue / context.getOutgoingEdgesNum();
    context.updateVertexValue(String.valueOf(newVertexValue)); }
/* Send new messages */
outgoingEdges = context.getOutgoingEdges();
while (outgoingEdges.hasNext()) {
        EdgeNode = outgoingEdges.next();
        msg = new BSPMessage(Integer.parseInt(EdgeNode.getVertexID()),
                        Double.toString(sendMsgValue).getBytes());
        context.send(msg); }
    return; }
}
```

Fig. 6. PageRank algorithm based on BS-BSP platform

```
public class SumCombiner extends Combiner {
    public BSPMessage combine(Iterator<BSPMessage> messages) {
        BSPMessage msg;
        double sum = 0.0;
        do {
            msg = messages.next();
            String tmpValue = new String(msg.getData());
            sum = sum + Double.parseDouble(tmpValue);
        } while (messages.hasNext());
        String newData = Double.toString(sum);
    msg = new BSPMessage(msg.getDstPartition(), msg.getDstVertexID(),
                newData.getBytes());
        return msg; }
}
```

Fig. 7. The *combiner* class for *PageRank* algorithm on BC-BSP platform

6.2   K-Means on Metric Data

In this subsection, we describe the basic idea about k-means clustering on ordinary multi-dimensional metric dataset on BC-BSP platform. Because the data structure of BC-BSP is designed for processing graph data, the multi-dimensional metric data must be converted in order to fit the input requirement of BC-BSP, but it is easy to do. So, we convert i^{th} data point (i.e. i^{th} line in data file) $<d_1, d_2, d_3,..., d_n>$ into the following format: $i:tagvalue$ $<tab>$ $1:d_1$ $2:d_2$ ... $n:d_n$. Where, the first part $<i:tagvalue>$ is regarded as a graph vertex, i is the line number of a record in the data file, and *tagvalue* can be used as the cluster tag that the data record belongs to and its initial value can be a random integer value. The second part $<tab>$ stands for *tab* key, maybe 4 or 8 blank-spaces. The others are the outgoing-edge-list constructed from the each dimensional value of the data record separated by a blank-space, the part before semicolon, such as 1, 2, stands for the outgoing-edge vertex ID and isn't used in the computing procedure; the part after semicolon, such as d_n, stands for the nth dimensional value d_n of the data point. The stop criterion may be the error of a cluster center point between the adjacent two super-steps is less than a given threshold. Therefore, user must design an *aggregator* to compute the error. But user need not write *Combiner* to combine messages since no message need be sent between each task in the adjacent two super-steps. The program for k-means algorithm on BC-BSP platform is omitted because the space limitation.

7   Experiments

Some experiments are done to evaluate the performance and extendibility of our system under different circumstances. Because Google's *Pregel* does not open source, we do not compare with it. We compare *PageRank* algorithm implemented on BC-BSP platform with the one based on MapReduce framework, and the ones on *Hama* and *Giraph*.

The hardware environments for the experiments are IBM shared-nothing cluster linked by Gigabit Ethernet, in which each node has 2 hyper-threaded 2.00 GHz Intel Xeon CPUs, 2GB memory, 73GB and 7200rpm hard disk. The experiments are done on Linux Redhat V5.6 and JDK1.6 for Linux.

7.1   Comparison between BC-BSP and MapReduce

Two kinds of datasets, real world data and synthetic data, are used in the experiments. The features of the datasets are listed in Table 1.

We use 9 nodes of the cluster to run our experiments, and setup the JVM memory to 1GB. The raw data are assigned to each node near-equally under BC-BSP platform. Under MapReduce framework, the number of *Mappers* is determined by *Hadoop* framework according to the data block size, and the number of *Reducers* is setup as 9 (i.e. 9 nodes). Based on the above environment and configuration, the performance comparison between the *PageRank* algorithm based on BC-BSP and MapReduce on real world data is in Fig. 8, and on synthetic data in Fig. 9.

The experiment results show that the performance of *PageRank* algorithm on BC-BSP platform is better than that on *MapReduce* framework when the data including graph data and the messages sent to other workers can be stored on the memory during the course of computing. While when the volume of data is relative large, where data including graph data and messages, the response time for *PageRank* algorithm on BC-BSP platform is large greatly than that on *MapReduce* because the former needs to use disk space as spilled space.

Table 1. The features of datasets for experiments

DatasetType	Dataset Name	#Vertex	#Edges	Data file size
Real world data	Wikipedia Talk network	2394385	5021410	45.4MB
	Autonomous system by Skitter	1,696,415	11095298	116MB
	Patent citation network	3,774,768	16,518,948	203 MB
	Live Journal social network	4,847,571	68,993,773	700 MB
Synthetic data	Syth1	10,000	676,640	4.5MB
	Syth2	50,000	7,543,140	56MB
	Syth3	100,000	21,136,232	160MB
	Syth4	4,000,000	53,991,808	685MB

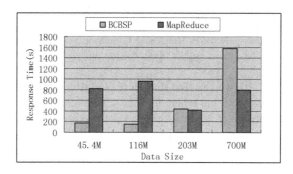

Fig. 8. Comparison between BCBSP and MapReduce on real dataset

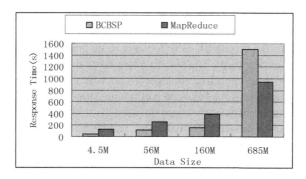

Fig. 9. Comparison between BCBSP and MapReduce on synthetic dataset

7.2   Comparison among BCBSP, Hama and Giraph

In order to test the processing ability among BC-BSP, Hama without disk help, and Giraph, we design an experiment to run *PageRank* algorithms on synthetic dataset. Because the difference of the processing capability of the three BSP-based system and the difference of their expressions on data, we generate the synthetic datasets with the same numbers of vertexes and edges, respectively. the experiments are executed on 1G JVM. The experimental results are shown in Fig. 10.

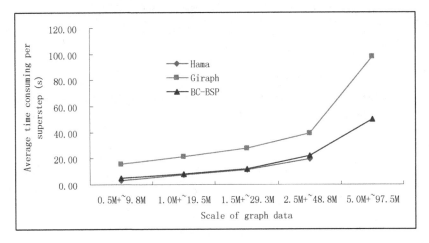

Fig. 10. Comparison among BCBSP, Hama and Giraph on another synthetic dataset, where 0.5M+~9.8M stands for the data set with 500000 vertexes and about 9800000 edges

By now, from Fig. 10, we can find that *Hama* is faster than BC-BSP a little more when the data scale is small, but when the data scale exceeds a certain value, such as 2.5M vertexes, the jobs on *Hama* platform can not execute because of memory over-flow, the average response time for a super-step on BC-BSP is faster some degree than that on *Giraph* on each test data set. We can also find that applications run on BC-BSP are faster than that on *Giraph*.

8   Conclusions and Discussion

This paper describes the system BC-BSP for large-scale graph processing based on BSP model under Java environment. The system implements the main functions mentioned in *Pregel*, and adds some optimized strategies to improve and enhance the performance of the system. It implements the balanced partitioning strategy in data partitioning stage in order to make each task have the approximately equal graph nodes to process. It im-plements disk swap function temporally to store graph data and the messages when they can not be hold in main memory to make the system can handle large-scale graph under constraint of computing and storage resources. We do many experiments to evaluate the performance of the system. We can conclude that the BSP-based applications have higher efficiency than that of *MapReduce*-based applications when the volume of data is relative not very large and it can be held in the memory during the course of processing; on the contrary the latter are better than the former. But our system can handle relative large-scale graph data when the computing and storage resources are limitation because it applies disk assistant mechanism.

Although we have done many efforts to optimize the system, there are many aspects to optimize and improve the system. For instance, a) we can optimize and improve the data structure for storage and presentation of graph data using template technique of Java to enhance the flexibility and to save the storage consumption; b) we should consider the locality and relevance of data at the data partitioning stage except considering the balance of each task.; c) we could enhance the capture and detection of every kinds of faults and handle each kind of faults using different policy; d) we could improve the programming skills to decrease memory consumption and increase the system performance. We can use object-pool technique to cache some kinds of object to decrease the CPU consumption on object constructions and the main memory occupation.

Acknowledgement. This work is partially supported by the Key National Natural Science Foundation of China under Grant No 61033007, the National Natural Science Foundation of China under Grant No. 61173028, and the joint Foundation of Ministry of Education of China and China Mobile under Grant No. MCM20125021.

References

1. Dean, J., Ghemawat, S.: MapReduce: Simplified Data Processing on Large Clusters. In: Proc. of 6th USENIX Symp. on Operating Syst. Design and Impl., pp. 137–150 (2004)
2. Brin, S., Page, L.: The anatomy of a large-scale hypertextual web search engine. Computer Networks and ISDN Systems 30(1-7) (1998)
3. Malewicz, G., Austern, M.H., Bik, A.J.C., Dehnert, J.C., Horn, I., Leiser, N., Czajkowski, G.: Pregel: A System for Large-Scale Graph Processing. SIGMOD (2010)
4. Welcome to Hama Project, http://incubator.apache.org/hama/
5. Snoek, J.: Computing PageRank using MapReduce. Technical Report, Report No. CSC2544. University of Toronto, Toronto (2008)
6. Ching, A., Kunz, C.: Giraph: Large-scale graph processing infrastructure on Hadoop, Hadoop Summit (2011)

DNA Encoding for Splice Site Prediction in Large DNA Sequence

A.T.M. Golam Bari[1], Mst. Rokeya Reaz[1], Ho-Jin Choi[2], and Byeong-Soo Jeong[1]

[1] Department of Computer Engineering, Kyung Hee University
1-Seocheon-dong, Gyeonggi-do, Yongin-si 446-701, Republic of Korea
[2] Computer Science Department, Korea Advanced Institute of Science and Technology
335 Guseong-dong, Yuseong-gu, Daejeon 305-701, Republic of Korea
{bari,rokeya,jeong}@khu.ac.kr,
hojinc@kaist.ac.kr

Abstract. Splice site prediction in the pre-mRNA is a very important task for understanding gene structure and its function. To predict splice sites, SVM (support vector machine) based classification technique is frequently used because of its classification accuracy. High classification accuracy of SVM largely depends on DNA encoding method for feature extraction of DNA sequences. However, existing encoding approaches do not reveal the characteristics of DNA sequence very well enough to provide as much information as DNA sequences have. In this paper, we propose new effective DNA encoding method which can give more information of DNA sequence. Our encoding method can provide density information of each nucleotide along with positional information and chemical property. Extensive performance study shows that our method can provide better performance than existing encoding methods based on several performance criteria such as classification accuracy, sensitivity, specificity and area under receiver operating characteristics curve (ROC).

Keywords: DNA sequence, gene prediction, nucleotide density, orthogonal encoding, support vector machine, splice site, ROC.

1 Introduction

As more whole-genome sequences are increasingly generated with the continued development of new high-throughput methods for DNA sequencing, gene identification becomes one of the important tasks for computational biology. In order to understand how the genome works, we need to identify a set of coding fragments, known as exons, which are separated by non-coding intervening fragments, known as introns. As shown in Figure 1, the boundaries between exons and introns are called splice sites. The vast majority of all splice sites are characterized by the presence of specific dimers: GT for donor and AG for acceptor sites. However, only about 0.1%~1% of all GT and AG occurrences in the genome represents true splicing sites. Thus accurate prediction of splice site is naturally required for a systematic study of eukaryotic genes.

B. Hong et al. (Eds.): DASFAA Workshops 2013, LNCS 7827, pp. 46–58, 2013.
© Springer-Verlag Berlin Heidelberg 2013

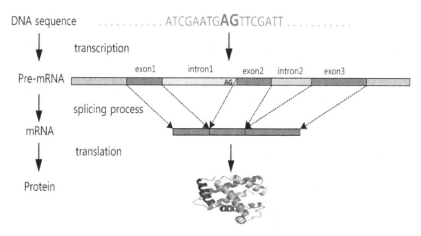

Fig. 1. Central Dogma and Splice Sites

For this reason, there have been a lot of research works for predicting gene's structure and its function. Several machine learning algorithms have been developed for splice site prediction such as Bayesian networks, ANN (artificial neural network), discriminant analysis, and SVMs. Among them, SVMs and related kernel methods are most frequently used for solving such problems [1-10] due to their high accuracy and capability to deal with high-dimensional large data sets.

When we use SVM based classification technique, the feature extraction is a very important step for better classification accuracy. For feature extraction, DNA encoding method has the advantage of simple process. It can also provide the characteristics of DNA sequences to transform splice site sequence to a feature vector. However, existing encoding approaches do not reveal the characteristics of DNA sequence very well enough to provide as much information as DNA sequences have. In this paper, we propose a new effective DNA encoding method which can give more information of DNA sequence. Our encoding method can provide density information of each nucleotide along with their positional information and chemical property. Extensive performance study shows that our method can provide better performance than existing encoding methods based on several performance criteria such as classification accuracy, precision, recall and ROC.

The paper is organized as follows: In Section 2, we briefly survey related work about splice site prediction. We describe our proposed encoding method in detail in Section 3. We explain experimental environment and analyze its results in Section 4. Finally Section 5 presents our conclusion.

2   Related Work

A large number of computational methods have been applied for solving biological sequence (DNA, Protein) analysis, finding gene regulation, protein protein interaction and so on. Splice site prediction is one of the important tasks to predict gene's structure and its function. Thus, there was a lot of research works to classify true splice sites effectively. Most of previous works are mainly based on HMM (Hidden Markov model), Neural Network, and several statistical analysis. Even though a large number of splice site prediction tools are publicly available, they still need to improve their performance for high prediction accuracy and capability of handling a large scale DNA sequences. Table 1 summarizes the characteristics of representative prediction tools [11].

Table 1. Expert systemsforsplicesiterecognition

Tools name	Organism for training data set	Learning model
GeneSplicer	Arabidopsis, Human	HMM + MDD
NETPLANTGENE	Arabidopsis	NN
NETGENE2	Human, Arabidopsis	NN + HMM
NNSPLICE0.9	Drosophila, human	NN
SPLICEDETECTOR	Arabidopsis,maize	Logit linear models
BCM-SPL	Human,Drosophila,yeast, plant,C.elegans	LDA
SPLICEVIEW	Eukaryotes	Score with consensus

* HMM – Hidden Markov Model, NN – Neural Network, MDD – Maximal Dependence Decomposition, LDA – Linear Discriminant Analysis.

SPLICEVIEW [12] searches for a match with a consensus sequence on a set of aligned functional sites considering the correlations between nucleotides of those sites. SPLICEDETECTOR [13] also uses the same approach as SPLICEVIEW does with some additional information. Positional weight matrix (PWM) determines the appearance probability of a given base at each position of the signal which can also be optimized by a neural network method, as proposed in NETPLANETGENE [14], NETGENE2 [15] and NNSPLICE [16].

On the other hand, SVM which is a powerful pattern recognition technique, is successfully applied for splice site prediction problem because of its high classification accuracy and capability of handling large-scale DNA sequences. In order to apply SVM, effective DNA encoding approach is necessary for transforming raw DNA sequences into vectors of feature space. On this account, several encoding methods are proposed and analyzed in many ways.

Salekdehet.et al. [2] proposed an encoding method which can consider the positional probability of each nucleotide while introducing another 4 ambiguous values to represent possibility of occurrence of some other nucleotides. However, it cannot distinguish distance value between ACC-AGG and ACC-ATT in the case of one matched and two unmatched nucleotides even though AGG is a more important sequence than ATT for splice site prediction.

In [1], for extracting more information from splice site sequences, they utilize three approaches, orthogonal encoding, codon usage, and sequential information. Huang et. al. [9] proposed four different encoding approaches and compare their performance. They are mono-nucleotide (MN) encoding which maps each of 4 DNA bases into an integer number, pairwise (PN) encoding which maps 16 possible pairs into an integer number, and combining frequency difference between the true and false sites (FDTF) encoding. Their experimentation indicates that PN with FDTF method produces the best accuracy. In [17], they classified 4 DNA bases as 4 different coordinates from the knowledge of biology which is based on nucleotide classification. Their experimental results show that the 4D representation provides good performance in measuring the evolutionary relationship among different species.

Zhang et. al [6] use weight matrix model for DNA encoding. The problem of this type of weight matrix is that it assumes each position is equally important and therefore each attribute (nucleotide) is independent. But attributes are not always independent and some positions may be essential while others may be trivial in the area of splice site prediction.

AKMA Baten.et al. [7] produced their best result in reduced Markov encoding where the conditional probability of a nucleotide at any location depends on its immediate predecessor. But the correlation between adjacent nucleotides does not reveal the global feature of splice sites. Markovian probability becomes complex and unrealistic when we consider high order Markov model for global feature.

However, none of the above methods consider density information of each nucleotide in DNA sequences which may be desirable information for splice site prediction. In this paper, we propose a new effective DNA encoding method which can give more information about DNA sequences. Our encoding method includes density information of each nucleotide along with positional information and chemical property. The extensive performance study shows that our method can provide better performance than existing encoding methods based on several performance criteria such as classification accuracy, sensitivity, specificity and area under ROC (auROC).

3   DNA Encoding

In order to accurately predict several functional sites (splice site, promotor site, translation site, etc.) in large DNA sequences, it is very important to extract appropriate features from DNA sequences. DNA encoding is one possible approach of representing features of DNA sequences. Nucleotide encoding in DNA sequences is a basic issue to improve the splice site prediction accuracy. Most of the available models using SVM for splice site prediction pay attention to nucleotide encoding for improving their classification models. As we discuss some nucleotide encoding techniques proposed by [1],[2],[6],[17] in Section 2, their methods are not enough to extract required features from DNA sequences. We need to consider not only statistical probability of nucleotide but also chemical property and distribution of nucleotide for representing features of DNA sequences more informatively. Another important requirements of DNA encodings are degeneracy and uniqueness to avoid information-loss. These two properties of DNA encoding are also helpful to identify the splice site.

We propose an encoding method based on chemical property of nucleotides and their density (distribution) information. Our encoding method also retains uniqueness and lack of degeneracy.

3.1 Nucleotide Density:

Let $\Sigma = \{A, T, C, G\}$ and $S = \{s_1, s_2, s_3, \dots, s_l\}$ is a DNA sequence of length l where $s_i \in \Sigma$ and i = 1, 2, 3, ... ,l. We further define that $|S|$ represents length of the string S and $|S_i|$ is the length of substring that starts at position 1 and ends at i. Then, the density d_i of any nucleotide s_i in the position i is formally derived as:

$$d_i = \frac{1}{|S_i|}\sum_{i=1}^{l} f(S_i) \text{ where } f(q) = \begin{cases} 1 & if\ s_i = q \\ 0 & otherwise \end{cases}, \ i = 1,2,3, \dots l \quad \text{and} \quad q \in \Sigma$$

The d_i acts as nucleotide's positional weight within a sequence. For example, consider a sequence ATAGTCATAA. The density of A is 1, 0.67, 0.43, 0.44, and 0.50, in the position 1, 3, 7, 9 and 10 respectively, T is 0.5, 0.40, and 0.37 in the position 2, 5 and 8 respectively, C is 0.17 in the position 6 and G is 0.25 in the position 4.

3.2 Nucleotide Chemical Property

Each nucleotide in DNA sequences may give different functioanlty according to its chemical structure. Thus, three coordinate values of our encoding scheme are determined by its chemical property of the nucleotide. First, Purine (A, G) and pyrimidine (C, T) both have rings – purines have two rings and pyrimidines have one. So, they will fall into same coordinate (here, x coordinate). Similarly amino (A, C) and keto (G, T) group fall into y coordinate because they have same functionality. Eventually, z coordinate is determined by strong or weak Hydrogen bond – strong H (C, G) and weak H (A, T) group.

We reduce S into a series of nodes $P_1, P_2, \dots, P_l$ whose coordinates $P_i = [x_i, y_i, z_i, d_i]$ (where i = 1, 2, 3, ... , l) satisfy the following equations:

$$xi = \begin{cases} 1 & if\ s_i \in \{A, G\} \\ 0 & if\ s_i \in \{C, T\} \end{cases} yi = \begin{cases} 1 & if\ s_i \in \{A, G\} \\ 0 & if\ s_i \in \{C, T\} \end{cases} zi = \begin{cases} 1 & if\ s_i \in \{A, G\} \\ 0 & if\ s_i \in \{C, T\} \end{cases}$$

As an example, the sequence "ATAGTCATAA" is represented by |1,1,1,1|0,0,1,0.5|1,1,1,0.67|1,0,0,0.25|0,0,1,0.4|0,1,0,0.17|1,1,1,0.43|0,0,1,0.37|1,1,1,0. 43|1, 1,1,0.50| where "|" represents a virtual separator and does not exist in real encoded data.

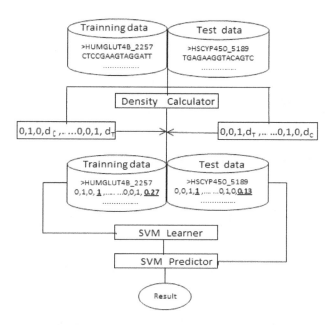

Fig. 2. Proposed Model

3.3   Proposed Prediction Model

Our prediction model use SVM for classifying true and false splice site as shown by Fig. 2. Each nucleotidesin a DNA sequence are converted into four dimensions in feature space. The first three are either 1 or 0 based on nucleotide chemical property and the last one is any floating point number between 0 and 1 as given by density(di). If there are n nucleotides in an acceptor or donor in the input space then we convert the data into 4n + 1 dimensions in the feature space. The last dimension value represents the class label (+1 or -1/ true or false) here. The feature space of our input data is simple and easy. It provides density information of each nucleotide in every fourth position.

We convert the dataset (DNA sequence) into feature vectors for SVM based on our encoding scheme. Firstly, we convert each nucleotide into its (x,y,z) coordinates. Secondly we calculate their density (d) in (x,y,z,d) dimensions. Finally, we concatenate their (x,y,z) plot with density. After that, we fed encoded training data into SVM. We name this phase 'SVM Learner' which is SMO algorithm implemented in Weka. RBF and polynomial kernels are chosen to train the SVM. After the finishing of training phase, 'SVM learner' is ready to predict the unknown data. That is why, this phase is named as 'SVM predictor' because it can now predict the unknown data based on its learning phase. So, we fed test data into SVM predictor. The test data is not fed into SVM learner because this data is only used for measuring the performance and stability of SVM. Finally, the predictor generates classification result.

4 Experimental Results

4.1 Experimental Environment, Dataset and Performance Metric

In order to evaluate the performance of our proposed encoding method, we have performed several experiments by using NN269 [23] dataset. Table 2 shows the characteristics of the dataset. These dataset are publicly available for analyzing splite site prediction model. The donor datasets have 7-bp (base pair) of the exon and 8-bp of the following intron (starting with GT).The acceptor data sets have 70-bp in the intron (ending with AG) and 2-bp of the following exon.

Table 2. Datasetcharacteristics

Dataset	Acceptor		Total	Donor		Total
	True	False		True	False	
Training	1116	4672	5788	1116	4140	5256
Testing	208	881	1089	208	782	990
Total	1324	5553	6877	1324	4922	6246

Since the existing Reduced MM1 - SVM outperforms the other methods like Information Content (IC Shapiro) and MM1- SVM, and MM1- SVM/WMM1 - SVM [8] outperforms Loi - Rajapakse [24], NNSplice [16] and GeneSplicer [25], we compare the performance of our algorithm with only Reduced MM1 - SVM [7]. At first, we show the overall performance of the proposed encoding using SVM kernels. We use polynomial and RBF kernel with different parameters. After that, we evaluate the performance of sparse encoding in the same experimental environment. Then we add density in sparse encoding and reevaluate the performance to show the importance of density in already existing model. Finally, we compare our classification accuracy with [7] for performance evaluation. Our programs were written in Python 2.7, and run with the Windows XP operating system on a Pentium dual-core 2.13 GHz CPU with 2 GB main memory. We used BioPython 1.60 for sequence encoding and SMO algorithm of Weka 3.7.5 [26] for SVM classification.

To evaluate the classification performance, we use several performance evaluation methods such as the sensitivity (S_n), specificity (S_p), accuracy, receiver operating characteristics curve (ROC), and the area under ROC as described in the following.

$$Sensitivity(Sn) = \frac{TP}{TP + FN} \times 100, \quad Specificity(Sp) = \frac{TN}{TN + TP} \times 100$$

$$Accuracy = \frac{TN + TP}{TN + TP + FN + FP} \times 100$$

Where TP, TN, FP, FN represents true positive, true negative, false positive and false negative respectively. Plotting Sn against 1-Sp produces the ROC [27]. ROC analysis is an effective and widely used for assessing the performance of classifiers. The larger the value of Sn, Sp, accuracy, ROC the more accurate the prediction is.

4.2   Efficiency of Our Model in SVM Kernels

We use 1324 true acceptors and 5553 false acceptors (total 6877 = 1324 + 5553) from NN269 dataset to evaluate the performance of the proposed model in SVM. 1116 out of 1324 true acceptors are training and rest 208 (1324 − 1116 = 208) true acceptors are for test purpose. In case of false acceptor, 4672 out of 5553 are training and rest 881 (5553 − 4672 = 881) are test. At first, we train our model with 5788 (1116 + 4672 = 5788) acceptors and we call this model 'SVM predictor' as described in Fig. 2. However, we merge these sites into a single training file and reallocate the true and false sites randomly in the file to protect the model being biased. Similarly, we prepare the test file with 1089 (881 + 208 = 1089) acceptors using the same technique of reallocation in training file to get the actual performance of the proposed model. Eventually, we reevaluate the performance of our method with 'SVM predictor'. In this step, the test data are fed into the predictor to evaluate the real performance of the proposed encoding method. Table 3 shows the performance of the model for acceptor and donor splice site data.

Table 3. Performanceevaluationof NN269 acceptoranddonorsplicesite

	TP	TN	FP	FN	Sn	Sp	Accuracy	auROC	Kernel
Donor	183	756	22	25	87.98	97.19	95.25	98.3	Poly
	185	758	24	23	88.94	96.93	95.25	98.24	RBF
Acceptor	161	856	25	47	77.40	97.16	93.39	97.9	Poly
	165	857	24	43	79.3	97.28	93.85	97.91	RBF

We used grid.py in LibSVM [28] to obtain the best kernel parameter value as well as tried some heuristic value for C, E and γ. Then we stored the parameters that produce best performance. 5-fold cross validation was used. For donor site, C = 0.001 and E = 2 are used for polynomial kernel and C=8, γ = 0.007 for RBF kernel. Similarly, we used C = 3, E=2.5 for acceptor when using polynomial kernel and C=8, γ = 0.004 for RBF kernel. The best values for TP, TN, FP, FN, Sn,Sp, accuracy and auROC are shown for both splice sites in Table 3.

Figure 3 and Figure 4 show the ROC curves for acceptor and donor splice sites respectively.   The ROC curves show the best classification performance for both kernels. For donor site, we get the best auROC value 98.30 which is 0.39 higher than the best ROC value of acceptor.

In the case of donor site, we follow the same reallocation process on training and test file as we do in acceptor site. The total number of donor sites is 6246. Of them, 5256 are training and 990 are for testing purpose. The training file contains 1116 true and 4140 false donor and the test file has 1324 true and 4922 false donor sites. The best classification performance for donor site is shown in Table 3.

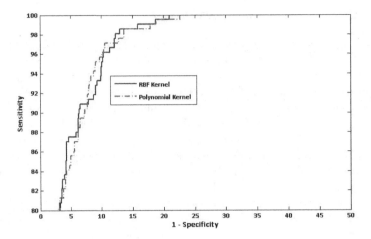

Fig. 3. ROC curves showing the classification performance for NN269 acceptor

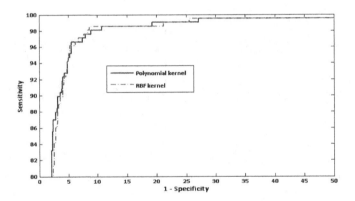

Fig. 4. ROC curves showing the classification performance for NN269 donor

4.3   Effectiveness of Density with other Encoding

In this section, firstly we implement sparse encoding (A->1000, C->0100, G->0010, T->0001) and evaluate the performance of this encoding on NN269 dataset for splice site prediction. Secondly, we add density as a fifth dimension of sparse encoding and reevaluate the performance. The entire file processing of training and test data are same as described in section 4.2. The dimension in feature space of a sequence with n nucleotide is 5n + 1 because sparse itself is a 4 bit encoding. The extra 1 in 5n + 1 is class label of splice site. Thirdly, we compare the performance between sparse encoding and density added sparse encoding to show the effectiveness of density on already established models. Eventually, we compare the performance between our proposed method and density added sparse encoding. Table 4 shows the performance of sparse encoding and density added sparse encoding.

Table 4. Classificationperformancewithand withoutdensity

Methods	Without density		With density		Splice Site
	Accuracy	auROC	Accuracy	auROC	
Proposed	-	-	93.85	97.91	Acceptor
Proposed	-	-	95.25	98.30	Donor
Sparse	93.57	97.7	93.74	97.8	Acceptor
Sparse	93.79	97.6	95.25	98.3	Donor

For sparse encoding, the best accuracy and auROC we get for acceptor are 93.57 and 97.7 respectively. In case of donor, the accuracy is 93.79 and auROC is 97.6. For density added sparse encoding, the accuracy and auROC are increased for acceptor. The increment of accuracy and auROC for donor splice site is remarkable. The accuracy and auROC for donor is 95.25 and 98.3 respectively. Our proposed method outperforms than sparse encoding and density added sparse encoding. So it is clear that the accuracy and auROC increase when density is added in sparse encoding. Fig. 5 shows the ROC curves for sparse encoding and density added sparse encoding. For acceptor, the curves differentiate in some places but eventually sparse with density curve outperforms than others.

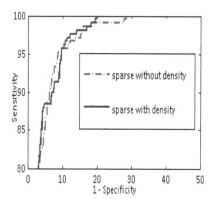 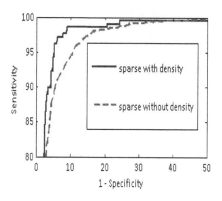

(a) ROC curves for NN269 acceptor splice site data applied in sparse and density added sparse encoding

(b) ROC curves for NN269 donor splice site data applied in sparse and density added sparse encoding

Fig. 5. ROC curves showing the classification performance of NN269 splice site data on sparse encoding and density added sparse encoding

As shown in Fig. 5 (b), the difference between sparse and density added sparse encoding is vividly shown. The difference of auROC is 0.7. It can be concluded that the performance of density added sparse encoding is significantly superior to that of sparse encoding.

4.4   Comparison of Classification Performance with Reduced MM1-SVM

In this experimental section, we compare the proposed model with reduced MM1-SVM to verify the practical applicability of the models obtained. Reduced MM1-SVM extracts sequence feature using first order Markov model where it generates some position specific probabilistic parameters or emission probabilities for input sequence data to learn the conserved sequence pattern at upstream and downstream regions surrounding the splice site motifs (GT-AG). Table 5 shows the classification performance comparison of our model with [7].

Table 5. auROCcomparisonofthe models

	Proposed SVM (Poly)	Proposed SVM (RBF)	Reduced MM1 (GRBF)	Reduced MM1 (Poly)	MM1 (Poly)	IC. Shapiro (Poly)
Acceptor	97.90	97.91	97.41	96.96	96.74	96.23
Donor	98.3	98.24	97.9	97.65	97.62	96.66

Compared to MM1-SVM [8], reduced MM1-SVM and IC. Shapiro from [7], our method gives a better performance. For acceptor sites, our model gives the best au-ROC 97.90 (97.91) which is 0.94 (0.50) higher than that of reduced MM1-SVM poly (reduced MM1-SVM GRBF), 1.16 higher than that of MM1-SVM poly and 1.67 higher than that of IC. Shapiro SVM poly. In case of donor sites, our model produces the best AUC 98.30 (98.24) which is 0.65 (0.34) higher than that of reduced MM1-SVM GRBF (reduced MM1-SVM poly), 0.68 higher than that of MM1-SVM poly and 1.64 higher than that of IC. Shapiro SVM poly (see Table 5).

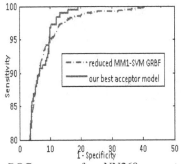

(a) ROC curves for NN269 acceptor splice site data applied in proposed method and reduced MM1-SVM GRBF

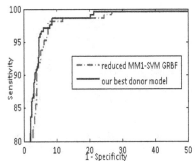

(b) ROC curves for NN269 donor splice site data applied in proposed method and reduced MM1-SVM GRBF

Fig. 6. ROC curves showing the comparison of performance between our model and reduced MM1-SVM GRBF using NN269 dataset

Fig. 6 shows the comparison of performance between our model and reduced MM1-SVM. As shown in both Fig. 5(a) and Fig. 5(b), our model is clearly the superior model for the identification of both acceptor and donor splice sites.

5   Conclusion

The accurate prediction of splice site is the key point of gene identification. Several mathematical models and encoding approaches are used to recognize the splice site. The accuracy of these approaches depends on their feature extraction method. Our simple and easy approach of density information largely increases the accuracy of the problem area. We consider the chemical property of nucleotides for encoding approaches. Our encoding approach with nucleotide density is easy to implement and simple but produce better result than others.

Acknowledgement. This work was supported by the National Research Foundation (NRF) grant (No. 2012-0001001) of the Ministry of Education, Science and Technology (MEST) of Korea.

References

1. Wei, D., Zhuang, W., Jiang, Q., Wei, Y.: A New Classification Method for Human Gene Splice Site Prediction. In: He, J., Liu, X., Krupinski, E.A., Xu, G. (eds.) HIS 2012. LNCS, vol. 7231, pp. 121–130. Springer, Heidelberg (2012)
2. Salekdeh, A., Wiese, K.: Improving splice-junctions classification employing a novel encoding schema and decision-tree. In: IEEE Congress on Evolutionary Computation, pp. 1302–1307 (2011)
3. Loris, N., Lumini, A.: Identifying Splice-Junction Sequences by Hierarchical Multi classifier. Pattern Recognition Letters 27(12), 1390–1396 (2006)
4. Nantasenamat, C., Naenna, T., Isarankura-Na-Ayudhya, T., Prachayasittikul, V.: Recognition of DNA Splice Junction Via Machine Learning Approaches. Experimental and Clinical Sciences International Online Journal for Advances in Science 4, 114–129 (2005)
5. Ying-Fei, S., Xiao-Dan, F., Yan-Da, L.: Identifying Splicing Sites in Eukaryotic RNA: Support Vector Machine Approach. Computers in Biology and Medicine 33(1), 17–29 (2003)
6. Ya, Z., Chao-Hsien, C., Yixin, C., Hongyuan, Z., Xiang, J.: Splice Site Prediction Using Support Vector Machines With a Bayes Kernel. Expert Systems with Applications 30(1), 73–81 (2006)
7. Baten, A., Halgamuge, S., Chang, B.: Fast Splice Site Detection Using Information Content and Feature Reduction. BMC Bioinformatics 8 (2008)
8. Baten, A., Halgamuge, S., Chang, B., Jason, L.: Splice Site Identification Using Probabilistic Parameters and SVM Classification. BMC Bioinformatics 7 (2006)
9. Huang, J., Li, T., Chen, K., Wu, J.: An Approach Of Encoding for Prediction of Splice Sites using SVM. Biochimie 88(7), 923–929 (2006)
10. Yifei, C., Feng, L., Vanschoenwinkel, B., Manderick, B.: Splice Site Prediction using Support Vector Machines with Context-Sensitive Kernel Functions. Journal of Universal Computer Science 15(13), 2528–2546 (2009)

11. Mathe, C., Marie-France, S., Schiex, T., Rouze, P.: Current Methods of Gene Prediction, Their Strengths and Weakness. Nucleic Acids Research 30(19), 4103–4117 (2002)
12. Rogozin, B., Milanesi, L.: Analysis of Donor Splice Signals in Different Eukaryotic Organisms. J. Mol. Evol. 45(1), 50–59 (1997)
13. Kleffe, J., Hermann, K., Vahrson, W., Wittig, B., Brendel, V.: Logitlinear Models for the Prediction of Splice Sites in Plant Rre-mRNA Sequences. Nucleic Acids Res. 24, 4709–4718 (1996)
14. Hebsgaard, S.M., Korning, P.G., Tolstrup, N., Engelbrecht, J., Rouzé, P., Brunak, S.: Splice Site Prediction in Arabidopsis Thaliana Pre-mRNA by Combining Local and Global Sequence Information. Nucleic Acids Res. 24, 3439–3452 (1996)
15. Tolstrup, N., Rouzé, P., Brunak, S.: A Branch Point Consensus from Arabidopsis Found by Non-circular Analysis Allows for Better Prediction of Acceptor Sites. Nucleic Acids Res. 25, 3159–3163 (1997)
16. Reese, M.G., Eeckman, F.H., Kulp, D., Haussler, D.: Improved Splice Site Detection in Genie. In: First Annual International Conference on Computational Molecular Biology (RECOMB), pp. 232–240. ACM Press, New York (1997)
17. Bo, L., Mingshu, T., Kequan, D.: A 4D Representation of DNA Sequences and Its Application. Chemical Physics Letters 402(4-6), 380–383 (2005)
18. Nafiseh, J., Iranmanesh, A.: A Novel Graphical and Numerical Representation for Analyzing DNA Sequences Based on Codons. MATCH Commun. Math. Comput. Chem. 68, 611–620 (2012)
19. Aram, V., Iranmanesh, A.: 3D-Dynamic Representation of DNA Sequences. MATCH Commun. Math. Comput. Chem. 67, 809–816 (2012)
20. Chi, R., Kequan, D.: Novel 4D Numerical Representation of DNA Sequences. Chemical Physics Letters 407, 63–67 (2005)
21. Liao, B., Li, R., Zhu, W., Xiang, X.: On the Similarity of DNA Primary Sequences Based on 5D Representation. Journal of Mathematical Chemistry 42, 47–57 (2007)
22. Liao, B., Tian-ming, W.: Analysis of Similarity/Dissimilarity of DNA Sequences Based on Nonoverlapping Triplets of Nucleotide Bases. Journal of Chemical Information and Modeling 44, 1666–1670 (2004)
23. Reese, M.G., Eeckman, F., Kupl, D., Haussler, D.: Improved Splice Site Detection in Genie. Journal of Computational Biology 4(3), 311–324 (1997)
24. Rajapakse, J.C., Loi, S.H.: Markov Encoding for Detecting Signals in Genomic Sequences. IEEE/ACM Transactions on Comutational Biology and Bioinformatics 2(2), 131–142 (2005)
25. Pertea, M., Lin, X., Salzberg, S.L.: GeneSplicer: A New Computational Method for Splice Site Prediction. Nucleic Acids Research 29(5), 1185–1190 (2001)
26. Hall, M., Eibe, F., Holmes, G., Pfahringer, B., Reutemann, P., Witten, I.H.: The WEKA Data Mining Software: An Update. SIGKDD Explorations 11(1), 10–18 (2009)
27. Fawcett, T.: ROC Graphs: Notes and Practical Considerations for Data Mining Researchers. Technical Report HPL -2003-2004, HP Laboratories, Palo Alto (2003)
28. Chih-Chung, C., Chih-Jen, L.: LIBSVM: A Library for Support Vector Machines. ACM Transactions on Intelligent Systems and Technology 2(3) (2011)

Using a Pipeline Approach to Build Data Cube for Large XML Data Streams

Hao Gui[1] and Mark Roantree[2,3]

[1] International School of Software, Wuhan University, China
hgui@whu.edu.cn
[2] School of Computing, Dublin City University
mark@computing.dcu.ie
[3] CLARITY: Centre for Sensor Web Technologies

Abstract. XML has become a widely used standard for data representation, distribution and sharing. The concept of the Sensor Web has led to web generated sensor data in many diverse applications where delivery of the sensed data takes place using the Web. In order to obtain useful knowledge from XML sensor data, data warehouse and OLAP applications aimed at providing support for decision making for operational data must be developed. In this paper, we present a pipeline design based OLAP data cube construction framework designated for real time web generated sensor data, transforming sensor data into XML streams conforming to an underlying data warehouse logical model, which constructs corresponding data cubes. As part of this work, we discuss how our cube construction and acceleration strategy improves the efficiency in managing large volumes of XML data.

1 Introduction

While XML is generally regarded as one of the most widely used standards for data representation, distribution and sharing, the emergence of the Sensor Web has resulted in a significant increase in volume and diversity in online XML data. The concept of the Sensor Web [9] has led to the online publishing of many new forms of sensed data. Applications vary from environmental analysis, power management, various forms of transport and sports event monitoring. The result is that large volumes of XML data are generated on a daily basis with significant potential for knowledge extraction from inside terabytes of XML data. While there is a long established method for OnLine Analytical Processing (OLAP) for relational data, similar efforts for XML data are not as well advanced.

1.1 Motivation and Goals

In our CityBikes project described in detail in §2, we monitor the bicycle rental patterns across a number of cities around the world. Each of these rental schemes provides a real time service for users but there is no work on exploiting this information to perform any form of analytics. What is required is a system that

B. Hong et al. (Eds.): DASFAA Workshops 2013, LNCS 7827, pp. 59–73, 2013.

can both manage the live data streams coming into the system and an underlying infrastructure to manage data volumes. This will necessitate the formation of multi-dimensional structures such as Cubes which support the data analysis, but these cubes must be created in real time and updated as new data arrives. In earlier work [7], we presented our architecture for harvesting data from rental schemes across multiple cities and described a system for optimizing queries across the XML data. However, this work did not facilitate either live streaming data, nor any form of complex data mining. Work in providing multi-dimensional analysis for XML cubes was introduced in [5], where we developed a metamodel for managing cube construction and maintenance for XML. As this work was based on the Dwarf approach [11], it provided a platform for Big Data applications as the the Dwarf approach has been re-evaluated for large datasets and improved since [4]. What we delivered in this work was a first attempt at constructing XML cubes using the Dwarf approach. While this ensured a scalable platform, our goals in this current paper were to introduce a method for streaming XML data based on this platform, which included a mechanism for continuous updating of XML cubes.

1.2   Contribution and Paper Structure

In this paper, we present a novel pipeline approach for OLAP data cube construction, designed for real time web generated sensor data. This approach facilitates the transformation of sensor data into XML streams that conform to the data warehouse logical model, builds the corresponding data cube tree and then serializes it into XML data cube representation. We have also provided an extensive evaluation which demonstrates the performance of our system.

This paper is structured as follows: in §2, we begin with an overview of the project and description of the Data Warehouse model; in §3, we provide an outline of the system architecture; in §4, we describe our system metamodel as it describes the constructs used to deliver our approach; in §5, we provide details of evaluation; in §6, we discuss some related research and finally in §7, we provide some conclusions.

2   Background and Data Warehouse Model

Many cities around the world now provide bicycle rental schemes where people rent (and return) bicycles from stations located throughout the city's center. Stations are equipped with sensors that monitor bike availability and publish such data to websites such as www.dublinbikes.ie. Consumers can connect to the website (either through PC or mobile application) to check where stations are, how many bikes are available for rent, how many spaces are available to return bikes, and what type of payment methods are available. This data is of great interest to both consumers and providers of the service. Consumers can check where to rent or return a bike while providers can understand at which station it is better to pick up or return bikes for maintenance in order to minimize service disruption. In effect, the web service offers an efficient mechanism for determining the current status of bike or space availability.

Our project analyzes bicycle rental transactions with data collected from each location at 30 second intervals. For the purpose of our experiments, we used a 3GByte dataset to test our management of these data streams. Irrespective of the city or service provider, each station consists of a number of parking stands for bicycles, the actual bicycles, and a sensor based system to determine the status of the parking stands (empty or occupied). For each station this information is made available through the service provider's web site which provides station ID, total number of bike stands, number of bikes available, number of free bike stands available, number of tickets and so on. In addition, maybe the bicycle rental statistic has some intrinsic relationship with other related factors, such as weather conditions, bus routes in the city, and even more. We also harvest weather information as this project is contributing to a larger, smart cities project.

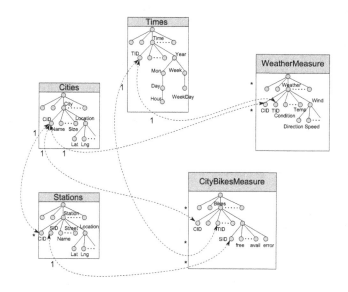

Fig. 1. Logical Model for the Data Warehouse

2.1 CityBikes Data Model

In Figure 1, the logical data warehouse structure of our CityBikes project is illustrated. A galaxy model is employed as the schema contains many fact tables with some common dimensions. In this figure, we have multiple fact sources: `CityBikesMeasure` and `WeatherMeasure`. If viewed from a relational data warehouse model perspective, the dimension and fact may contain an internal snowflake model. This is because both dimension data and fact data are in XML format, which may contain nested information. Therefore, if we flatten all hierarchical data into relations and draw the logical data warehouse model again, it will appear far more complicated than what is presented in Figure 1. Our requirements are to collect multiple Web sensor streams (which in this case are the weather information stream and citybikes information stream) from different Web sensor data providers. All are encoded in XML format, and the goal is to construct iceberg or standard multi-dimensional data cubes on the fly, which can subsequently be serialized into XML format for more complex data analysis. To retain the high level of interoperability provided by XML, together with its extensible nature, we do not wish to transform XML data into relations. Thus, we must complete data processing, storage, retrieval and management in the native XML format.

Furthermore, this is only the logical data warehouse model (or data warehouse definition), which describes nothing regarding OLAP related requirements nor the corresponding data cube necessary for construction. According to this DW definition, it is simple to build multidimensional data cubes for data analysis on either the `WeatherMeasure` or `CityBikesMeasure` fact separately. However, in our experimental CityBikes application, our goal is to carry out data analysis and data mining work on citybikes statistics with weather information as a form of *dynamic* information dimension in the same manner as static information dimensions (such as the `City` and `CityStation` dimension), although the sampling frequencies for CityBikesMeasure (perhaps several times per hour) and WeatherMeasure (perhaps several times per day) are totally different. The combination of `City` and `Times` facts can be used to build relationships for these two facts as shown in Figure 1, to provide the necessary functionality to analyze the weather information in the newly enriched citybikes data.

3   XDC Pipeline Architecture

In [3], the authors proposed a framework for multidimensional analysis of XML warehouses, which relied heavily on the usage of XPath and XQuery as its data extraction method. While equipped with the appropriate index for fact data retrieval, it still turns out to be very inefficient when it comes to data cube construction for large volumes of data. The authors in [13] presented a simple XML data cube model with an entire set of SQL-like OLAP Query expressions for their XML data cube, which help to describe the cube itself but must be transformed into large amount of atomic XQuery and XPath query plans in order to calculate the actual data cube. Both of these research efforts used XML documents to represent the facts and dimensions within an XML data warehouse, to logically model an XML data warehouse in a flexible way. We set out to develop a new cube construction method for Web generated sensor data in XML format using some of these ideas together with our XML Data Cube (XDC) framework presented in [5].

We begin with a brief overview of the XDC streaming framework. In Figure 2, we illustrate a data cube construction framework that uses a pipeline method to calculate data cubes directly from real time sensed XML data, without the requirement for executing XPath or XQuery plans by an underlying

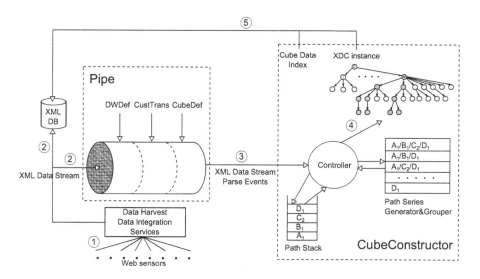

Fig. 2. XDC Stream Cube Framework

XML database. Source sensor data are supplied in the form of streams of XML events, referred to as source streams. Our pipeline design, which allows transformations to have a streaming characteristic, can transform one or more source XML streams into a single result XML stream that can be used as the input for data cube construction process. The result is that it does not need to build a tree representing the source data in memory, as this source data could potentially be endless, for certain sensor networks. The Result Events are generated as soon as Source Events appear and are processed.

In Figure 2, there are a number of components used to mange the sensor streams and construct the XML cubes required for analytics.

1. **Data Harvest** component. Reads and integrates sensor streams.
2. **Pipeline** component. Uses the Pipeline construct to perform a number of transformations (described in the next section).
3. **Cube Construction** component. Uses the XDC metamodel to build the XML cubes by extracting fact measurements and other static dimensional information from the source data.
4. **Storage** component. Provides persistence for both source XML data, cube definitions and the materialized cubes.

3.1 Pipeline Transformations

In this section, we describe the role of the Pipeline and the different types of transformations necessary to build cubes from XML streams. For the three phases, is is assumed that the DW definition is relatively stable, although changes can be made in phases two and three to build the data cube as required.

Phase 1: Transformation with DW Definition. The goal of Phase 1 is to convert the incoming XML streams into the logical Data Warehouse model. The DW definition includes the high level description of all dimension and fact data along with the relationships between them (as shown in Figure 1), whereas any schema information for the XML streams simply focuses on the low level logical and physical structure of the XML data. For example, facts such as the Weather measure and CityBikes measure may require a look-up and join process to deliver the transformations.

Phase 2: Customized Data Reduction or Transformation. The goal of Phase 2 is to carry out a number of pre-processing operations necessary for the cube construction process. These operations include, for example, elimination of simple semantic or syntactic heterogeneities, or data reduction operations. For example in our CityBikes Project, some analyzes do not require specific weather conditions (such as "Fair", "Thunder", "Partly Cloudy" and so on), and thus, we define several corresponding reduction rules (some conditions generalized into "Good", others into "Bad") to reduce the cardinality for a given dimension attribute with no impact on data analysis for that application. These reduction operations become acutely important in Big Data applications. This phase is also used to implement unit conversion strategies as our weather information is collected from different providers using different metrics. With similar operations, it will not only help to simplify the construction process and reduce the size of data cube, but can also customize the cube data for a specific data analysis and even data mining tasks. However, this phase does not concern itself with any aspects of cube construction.

Phase 3: Transformation with CubeDef. The goal of Phase 3 is to construct the final XML cube using the XDC metamodel first introduced in [5]. The cube definition includes the most critical parts for the cube construction process: dimensions, concept hierarchy within dimensions, interesting measurements for fact data, and types of aggregate functions (permissible by the Cube Model). Data not required for the construction of the current cube is eliminated, and any structural incompatibilities are adjusted. For example, in our CityBikes Project, what we want is to construct an XML data cube to enable analysis on usage statistics (bicycle usage pattern and status, to be exact) with respect to location information, city information, station information, time (year, month, day, hour) and weather status. It is necessary to join weather information into citybikes measurements in order to make it a dimension rather than a fact. Furthermore, the order of dimensions and the concept hierarchies must be determined in order to optimize the structure for cube construction. In our cube implementation, there are two concept hierarchies: the first being city and station; and the second being time hierarchy, i.e. year, month, day, hour. The pipeline will ensure a strict orchestration of XML data which may result in a small inflation, managed by the hierarchical nature of XML. However, it ensures the appropriate structure for the cube construction process.

After the final phase, the output XML stream will be fed into the CubeConstructor, which is discussed in the following section.

4   Cube Construction and Path Grouping

In this section, we describe our path grouping mechanism which is used to process incoming streams. We will begin however, with a brief overview of the XML Data Cube (XDC) metamodel first presented in [5]. This provides the platform on which our processing of XML streams can take place.

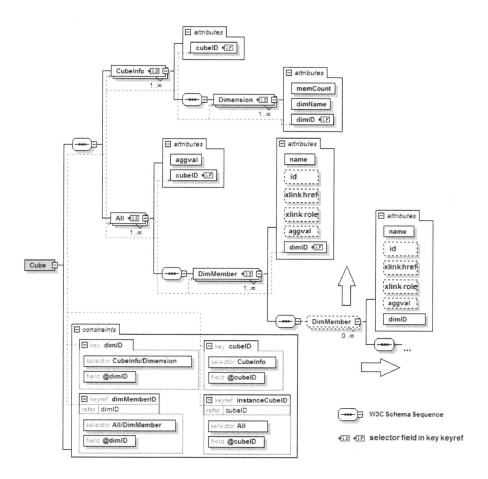

Fig. 3. XDC Metamodel

4.1 XML Data Cube Metamodel

The metamodel illustrated in Figure 4 has three major components which are used to manage the cube construction process.

- **Cubeinfo.** This element provides cube metadata together with dimension information such as name, id etc. As the size of the metadata is relatively small, it incurs little storage overhead.
- **All.** The All element contains the complete instance specific aggregation data of all cuboids composing the current cube lattice. Attributes id and aggval are used to for identification and measurement data with respect to each specific cuboid cell. If there is more than one aggregation value from measurement (such as sum and count), it is intuitive to turn aggval into a set of named attributes, even sub-elements, with structures to hold them. The DimMember element is self-contained and recursive from two directions, i.e. horizontally and vertically. For horizontal nesting, figure 4 shows that the child DimMember inside a parent DimMember can have any number of instances. In this way, DimMember can be used to capture the one to many logic in many XML datasets. The ellipsis notation in figure 4 illustrates vertical nesting: DimMember can be nested inside DimMember to arbitrary depth, and thus, used to represent the measurement data from a specific cuboid with specific series of dimensions.
- **Constraints.** The dash lines with arrows represent constraints inside the data cube structure. They model the relationships between dimensional metadata and instance data using XML Schema key/keyref mechanisms to ensure the integrity of the entire cube.

4.2 Data Cube Construction Process

In Figure 4, the forest on the left hand side provides a simple demonstration of the input XML stream containing bicycle usage after the three phase conversion process. This data stream contains source information intended for aggregation. For simplicity in this demonstration, it is assumed that each dimension (three dimensions in total in this example) only has two different possible values. In the input forest, any nodes at any level can repeat arbitrary times providing there is no corruption in corresponding dimension structures. It is important when designing cube structure that some dimensions are deliberately neglected for some data analysis tasks, so there exists many duplications with respect to remaining dimensions.

As part of this process, we reuse our Dwarf Derived Tree (DDT) construct from [5] which provides a gateway from a traditional Dwarf structure to an XML Dwarf structure. The corresponding DDT tree is shown on the right-hand side of Figure 4 which contains gray nodes (Data Nodes) and white nodes (Link Nodes). Data Nodes form the basic structure while simultaneously hosting cube data for several cube dimensions (namely City, City-Station and City-Station-Time). Each Data Node in the DDT contains four types of metadata: parent-child links in the tree structure; redundancy links as used in traditional Dwarf

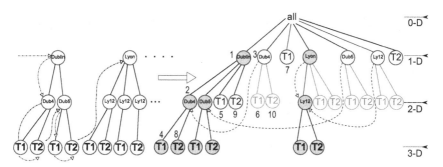

Fig. 4. XDC construction

model; attribute values for related dimensions; and aggregation data for corresponding cube cells. A Link Node also contains the first three types of metadata but in place of aggregation data, it contains view links consisting of all the Data Nodes in its set. In other words, Data Nodes provide aggregation data for data cubes, while Link Nodes offer viewing angles from different aspects. Nodes in semi-opaque format in figure 4 are those that are eliminated as a result of suffix redundancy [11].

The dashed lines in figure 4 link the first XML elements in order (left-hand side input stream): Dublin->Dub4->T1->T2->Dub5->T1->T2->Lyon. In the DDT tree on the right, the creation order for those XML elements is shown. Each Data Node has a set of observer Link Nodes. Thus, the observer Link Nodes for Dublin->Dub4->T1 include Dublin->T1, Dub4->T1 and T1, while Dublin->T1 is the observer of both Dublin->Dub4->T1 and Dublin->Dub5->T1. When the Cube process detects stream data items for the first time, it creates the corresponding Data Node, and then traverses the DDT to locate all observer Link Nodes and build the relationships between them. At this point, the algorithm also checks for redundancy and if located, ignores the entire subtree for that node and creates a redundancy link. If any node violates the condition of suffix redundancy [11], the redundancy link is deleted and the view links used instead.

4.3 Path Grouping Method

If we return to Figure 2, the CubeConstructor processes XML stream events one by one, calculates the aggregation values for related Data Nodes and produces related views from all kinds of multidimensional aspects with Link Nodes. There are several key components and internal data structures inside CubeConstructor, among them Controller is used to control the overall processing logic within CubeConstructor, Path stack, Path Series Generator and Grouper provide critical facilities to manipulate the cube tree effectively and efficiently during the entire construction process. The Path Grouping mechanism operates after generating the observers for a given Data Node, and Algorithm 1 shows the usage of our path grouping strategy when managing Data Nodes.

Algorithm 1. Cube Construction using Path Grouping Strategy

Input: a DataNode(n) which does not exist up to now

1.   Instantiate an empty stack structure(pathstack) for prefix path grouping

2.   startnode = parent of n

3.   while (startnode is not Node "All") {

4.   copy the attribute value of startnode into tempval

5.   startnode = parent of startnode;

6.   Instantiate a list of stack(stacklist) used to hold all paths of LinkNodes

7.   stacklist = PathGenerateFromStack(pathstack)

8.   outer for:

9.   for(each stack in stacklist){

10.   clone stack and insert attribute val of n to the bottom

11.   Node tmpnode = null;

12.   Node link = startnode;

13.   while(stack.size()¿0){

14.   tmpnode = link;

15.   if(link does not exist) continue outer for;

16.   val = stack.pop();

17.   link = link's child with attribute value val;

18.   }

19.   if(link does not exist){

20.   create a new LinkNode, and build up its link

21.   }

22.   else {update links of link}

23.   } //end for

24.   pathstack.push(tempval);

25.   } //end while

From Algorithm 1, rather than placing all LinkNode paths into a flat list, within the outermost while loop, this algorithm moves one dimension level upwards each time, and calculates all possible LinkNode paths for current situation. In other words, it groups all LinkNode paths into a hierarchical structures with respect to the current path prefix. This proves to be a very efficient way to create nodes and build links with significant elimination of tree traversing especially for large cube tree instances.

Assume the algorithm is processing a Data Node of level 4 with path (D_1, D_2, D_3, D_4), then all paths of the Link Nodes can be determined: $\{(D_1, D_2, D_4), (D_1, D_3, D_4), (D_2, D_3, D_4), (D_1, D_4), (D_2, D_4), (D_3, D_4), (D_4)\}$. In Algorithm 1, it will group it into several groups according to their prefix: Group1=$\{(D_1, D_2, D_4)\}$,
Group2=$\{(D_1, D_3, D_4), (D_1, D_4)\}$,
Group3=$\{(D_2, D_3, D_4), (D_2, D_4)\}$,
Group4=$\{(D_3, D_4), (D_4)\}$,

and when moving up through each level, the process finds the appropriate group and executes the corresponding operations (create Nodes and links). The cardinality of groups is much less than the cardinality of the original set, and the Link Nodes within each group comes from the same subtree. This form of optimization is increased as the numbers of dimensions grow.

Finally, the incorporation of the concept hierarchy into our framework is facilitated by a tweak in line 20 in Algorithm 1. Each time it begins the creation a new LinkNode, it first checks conformity with respect to the existing concept hierarchy definitions, a process which is outside the scope of this paper and part of current work.

5   Evaluation

The experimental dataset consists of a 6-month segment of bicycle sensor data with a total size of almost 3GByes, including over 15,000,000 bicycle usage transactions. All experiments were performed on an Intel Core2 E8400 PC clocked at 3.0GHz and with 4GB of memory. SAX parsers used in our programs are the default one shipped with JDK1.7.0, and XML data handling tasks are managed by Saxon9 open-source implementation version 9.3.0.5.

5.1   Cube Construction Performance

There are five dimensions and two concept hierarchies in this experiment and we elected to cut it from different angles. Our approach was as follows:

- In the legend in figure 5, for Scale (measure), we cut the original dataset by bike measurements, because the bikes usage sampling frequency is 30 times per hour, so cutting from this aspect will obviously reduce the size, without any significant effect on the structure of input data.
- For legend Scale(time) and Scale(city), we cut the original dataset by dimensions hour and city and thus, it affected both size and structure in different ways, and the performance curves followed the same increasing pattern when the size of datasets increased for each parameter. As three Scale parameters were used in the same dataset at value 15 on the X-axis, three curves converged at this point. In addition, reduction in the city dimension resulted in comparatively large change in structure, with the effect of the performance of Scale(city) being slightly better than that of Scale(time).

Two further experiments were carried out as part of our evaluation. In the city dimension, there is a concept hierarchy consisting of city and station, where some cities have large numbers of stations (for example, Bruxelles has almost 180 stations). In the first test, we deleted one of the members in this hierarchy (station) to examine the impact on data cube construction. As the cardinalities of the station members vary in size and are very large overall, there was a significant improvement in performance. This is illustrated in figure 5 by the **HierMemReduction** plot.

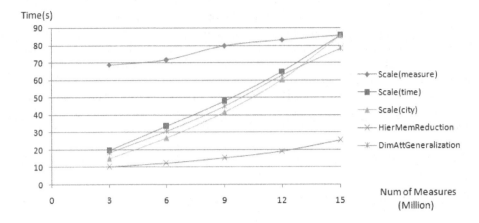

Fig. 5. Cube Construction of Citybikes Dataset

In the second of these experiments, the **weather** dimension was chosen to implement generalization. We imposed a more strict discretion rule to reduce the weather condition into either Good or Bad rather than its original values which in some cases had a quite broad range of value eg. "Fair", "Thunder", "Partly Cloudy" etc. This decision was made so that attributes such weather (40 in total across all city providers), do not make a significant contribution to the analysis of bicycle usage statistics. The result shown in figure 5, brought some performance improvements. However, it appears that the effects were far less than the removal of the concept hierarchy member **station**. The reason is that the cardinality of the **weather** dimension is large but the actual value distribution of this dimension is very sparse, as the weather condition changes little for a given city, even over an entire day.

5.2   Cube Construction Optimization

As discussed in Section 3, a number of optimization methods were developed as part of the cube construction process: the path stack, path grouper and hash table. We also performed a number of experiments to demonstrate the improvements the algorithms using these constructs delivered.

The test datasets for both experiments are identical with 3, 6, 9, 12 and 15 Million bicycle transaction measurements and cut at the hour point in the time concept hierarchy. The data series of legend `Scale(time)Optimized` represents the optimized run and its efficiency is far superior to the non-optimized run. In addition, the legend `NonOptimized` proves to be more sensitive to the growth of input XML sensor data. In other words, the gap between the optimization and non-optimization strategies widens as data volumes increase.

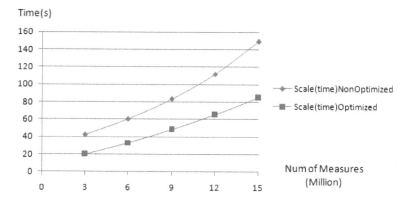

Fig. 6. Performance improvement in cube construction process

5.3   Eager Construction Optimization

In Figure 7, the legend CTime represents the total time cost for generating the required percentage of total nodes with eager construction. There is considerable benefit to eager constructions in that all aggregate values are precomputed, resulting in far quicker query processing times [12]. The legend PTime is the actual parsing time with the corresponding node intervals created in advance. The TTime legend is the total construction time for generating Link Nodes in advance and parsing the entire XML data stream. In effect, it is the summation of both CTime and PTime for a *specific* percentage.

In the CityBikes Project, roughly 90% of the dimensional information is static and not constantly changing with respect to fact measurements. Thus, these dimensions are candidates for eager construction and improve the overall cube construction performance. Link Nodes for these dimensions are populated as

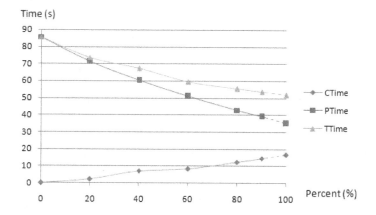

Fig. 7. Performance gain with Eager Construction

soon as possible using basic business logic, rather than being triggered by the actual data item. This is one reason why we place more static dimensions such as Location, City, Station first; with less static dimensions such as year, month, day, and hour in the middle; and finally, dynamic dimensions such as weather information last, when determining the appropriate order for all dimensions. It is only by using this approach that eager construction can be implemented in a controlled and gradual way with respect to size of the cube tree.

In Figure 7, the dash lines for each legend for the final 10% of the curve, represents the nodes coming from the (dynamic) weather dimension that we were unable to generate in advance. Those parts of lines represent a *theoretical* extension of the curve, and so dash lines are used as indicators only. However, it is clear from the results that eager construction boosts cube construction in our approach.

6   Related Research

With respect to the relational data model, Knowledge Discovery in Databases (KDD) has been studied for more than a decade. Indeed, there are many commercial or open source OLAP and data mining systems and platforms, that can be used to accomplish knowledge discovery tasks. However, the quite significant difference between the relational and XML models lies in the underlying data model, making it impossible to directly use or indiscriminately copy research from relational data warehouses and deploy in XML data warehouses [10].

Data warehousing and Online Analytical Processing (OLAP) technologies are now moving onto handling complex data that mostly originates from the web[1], much like the Web generated sensor data used in our research. One approach to an XML data warehouse design methodology was introduced in [8] where they constructed XML Cubes from the schema. However, this work did not tackle the issues of cube reduction in the way necessary to to manage Big Data unlike the Dwarf approach on which our work is based.

Some efforts have focused on using a relational approach to managing XML streams. However, the difficulty in converting between both formats strongly motivates the need for a purely XML approach as shown in a survey of these approaches [2]. In this work, the author examines each approach, detailing the open issues which strongly mitigate against this type of solution for a Big Data approach.

Finally, in [13], the authors present an XML data cube representation, which is not powerful enough to capture arbitrary multidimensional data. They require a customized SQL-like query language XRQL, to implement GROUP BY and other aggregate functions to support typical OLAP operations. This work will not suit high volumes of XML data, especially where cubes much be constructed on the fly.

7   Conclusions

In order to provide for applications such as data mining and analytics involved in Big Data applications, solutions similar to the Dwarf model are needed as it is proven to manage very high data volumes efficiently, together with highly optimized storage capabilities. In this paper, we present our approach to Big Data streaming management based on the Dwarf approach. Our framework optimizes incoming streams and builds highly efficient cubes on the fly. In our evaluation, we demonstrated the value of our framework using separate experiments for both optimized and non-optimized approaches.

References

1. Boussaid, O., Darmont, J., Bentayeb, F., Loudcher, S.: Warehousing complex data from the web. Int. J. Web Eng. Technol. 4, 408–433 (2008)
2. Cuzzocrea, A.: Cubing Algorithms for XML Data. In: Proceedings of DEXA Workshops, pp. 407–411. IEEE Computer Society (2009)
3. Park, B.-K., Han, H., Song, I.-Y.: XML-OLAP: A Multidimensional Analysis Framework for XML Warehouses. In: Tjoa, A.M., Trujillo, J. (eds.) DaWaK 2005. LNCS, vol. 3589, pp. 32–42. Springer, Heidelberg (2005)
4. Dittrich, J., Blunschi, L., Salles, M.: Dwarfs in the Rearview Mirror: How Big are they Really? Proceedings of VLDB Endowment 1, 1586–1597 (2008)
5. Gui, H., Roantree, M.: A Data Cube Model for Analysis of High Volumes of Ambient Data. Procedia Computer Science 10, 94–101 (2012)
6. Marks, G., Roantree, M., Murphy, J.: Classification of Index Partitions to Boost XML Query Performance. In: Parsons, J., Saeki, M., Shoval, P., Woo, C., Wand, Y. (eds.) ER 2010. LNCS, vol. 6412, pp. 405–418. Springer, Heidelberg (2010)
7. Marks, G., Roantree, M., Smyth, D.: Optimizing Queries for Web Generated Sensor Data. In: 22nd Australasian Database Conference (ADC 2011), Perth, Australia (2011)
8. Parimala, N., Pahwa, P.: From XML Schema to Cube. International Journal of Computer Theory and Engineering 1(3), 236–243 (2009)
9. Roantree, M., Sallinen, M.: The Sensor Web - Bridging the Physical-Digital Divide. ERCIM News 76 (2009)
10. Rusu, I.L., Wenny, R., David, T.: Partitioning methods for multi-version XML data warehouses. Distributed Parallel Databases 25, 47–69 (2009)
11. Sismanis, Y., Deligiannakis, A., Roussopoulos, N., Kotidis, Y.: Dwarf: Shrinking the PetaCube. In: Proceedings of ACM SIGMOD, pp. 464–475 (2002)
12. Widom, J.: Research Problems in Data Warehousing. In: Proceedings of the International Conference on Information and Knowledge Management (CIKM), pp. 25–30. ACM (1995)
13. Ykhlef, M.: On-Line Analytical Processing Queries for eXtensible Mark-up Language. Information Technology Journal 8(4), 521–528 (2009)

A Study on MapReduce Processing for Multi-dimensional Continuous Query

Doseong Jeong, Seungwoo Jeon, and Bonghee Hong

Dept. of Computer Engineering
Pusan National University, South Korea
{saladin6869,i2825t,bhhong}@pusan.ac.kr

Abstract. A huge volume of sensor stream data can be efficiently handle with the MapReduce framework for processing multi-dimensional continuous queries. The MapReduce originally has been used for batch processing, not real-time querying. In this paper, we propose a new idea of transforming query regions of multi-dimensional continuous queries into multiple key values. At the Map stage, key-value pairs of input data stream would be mapped into CQ-based key values that would be also grouped by the same continuous query in the Reduce stage.

1 Introduction

A large number of sensor nodes can be used for collecting temperature, humidity, GPS data to monitor a real application environment. Data Stream, consisting of data elements continuously produced in sequence, is the collection of real-time, continuous, fast and infinite data. In terms of dealing with queries about data stream, however, it is difficult to manage them using SQL because the amount of the data can be huge in proportion to the number of sensor devices. A large number of continuous queries should be efficiently executed on a big stream data generated every regular cycle. The data stream will be temporarily buffered before being handled so that it can be renewed if there is new stream data to be arrived during the buffering time.

In order to solve the bottleneck of performance of processing continuous queries, we consider to use distributed parallel processing, like MapReduce[1][2]. The MapReduce has been developed for batch processing of big data. The Map Reduce Online[3] is a modified version of the MapReduce that adds new abilities for real-time processing. It supports real-time processing by pipelining some part of the data that are already treated in the Map stage into the Reduce stage.

This paper discovers the problem of using the MapReduce online for supporting continuous queries over continuously incoming stream data. To apply the MapReduce framework on continuous query processing, it is necessary to define a suitable key for Map task. For a number of continuous queries, there are many overlapped query regions to represent specific query conditions. The key idea in this paper is to transform incoming sensor data into key values which correspond to specific query regions.

B. Hong et al. (Eds.): DASFAA Workshops 2013, LNCS 7827, pp. 74–78, 2013.

For incoming sensor data stream, we allocate dynamically partitioned data stream into a number of map tasks where input data would be converted into multiple key values for overlapped query regions. The next stage is to collect and sort key values before going into the Reduce stage. Finally the related key values are to be merged for corresponding continuous queries at the Reduce stage.

Section 2 mentions related works. Section 3 covers the design of map tasks for processing multi-dimensional continuous query and the procedure of pipelined processing at the Reduce stage. Finally, We summarize key ideas in Section 4.

2   Related Work

2.1   MapReduce Online

Hadoop Online Prototype(HOP)[3] includes to support pipelining behaviors, while preserving the full-featured MapReduce framework. This provides early returns on long-running jobs via online aggregation, and continuous queries over streaming data. HOP benefits are demonstrated by pipelining both within and across MapReduce jobs.

2.2   Continuous Queries

MapReduce can be considered to analyze streams of constantly-arriving data, such as URL access logs and system console logs. MapReduce is largely done in batch mode that only provides periodic views of activities. The MapReduce can bring significant delay in a data analysis process that ideally should run in near-real time. The pipelined version of Hadoop allows near-real-time analysis of data streams. However the pipelined version didn't mention the issues of parallel processing of continuous queries[4][5][6][7].

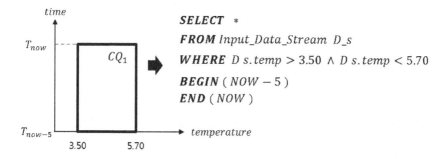

Fig. 1. Representation of two-dimensional continuous queries

3 MapReduce Design

3.1 Multi-dimensional Continuous Query

Figure 1 shows the representation of two-dimensional continuous queries. Temperature value is shown by the X axis, and time value is shown by the Y axis. For example, CQ1 is interested in temperature between 3.50 and 5.70. Figure 1 at the right side shows an example of continuous query specification.

3.2 MapReduce Process Flow

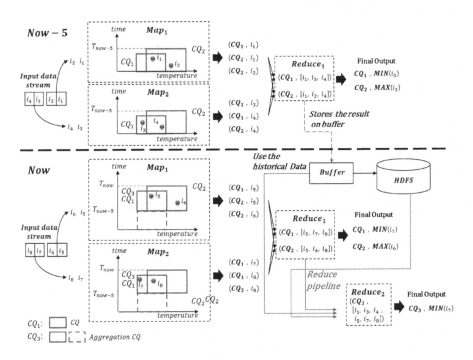

Fig. 2. A method of transforming query regions of CQ

Figure 2 shows a method of transforming overlapped query regions of multi-dimensional continuous queries into multiple key values. Input data stream are split into several map tasks that has the same processing routine. Intermediate result is based on key-value pairs where the key is the id of CQ and the value is stream data. In the next stage, all of the data are grouped by the same CQ-key value. Let us assume CQ3 can be processed by using the aggregation result of CQ1. To provide the result of CQ3 with that of CQ1, we need to use a pipelined grouping of related CQ3 at the Reduce stage.

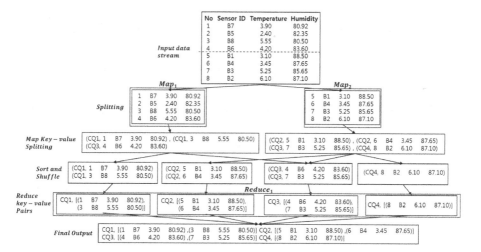

Fig. 3. MapReduce Process Flow

Figure 3 illustrates showing the whole processing flow of MapReduce from input to output. InputStream in the figure shows examples of collected sensor data. We assume four conditions queries are already registered and two map tasks are defined. Each map task transforms input stream data into key-value pairs. For each map task, it is necessary to support continuous query index to speed up transforming input data stream into key values. Later on, specific continuous queries can be added and/or deleted. To support dynamically changed CQ, we need to support updating of CQ index.

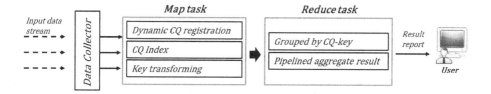

Fig. 4. MapReduce Processing

After sorting and shuffling of key values, the same key values for a specific continuous query should be grouped by each key value at the Reduce stage. The other problem is to make an aggregate result over a given time interval. we further continue to study on generating a pipelined aggregate result at the Reduce stage. Finally, the result of the sensor data stream that meet the conditions specified in a continuous query will stored HDFS.

The processing of continuous queries can be achieved by extension of the MapReduce framework shown in Figure 4. The Map task should include functions of dynamic registration of CQ, CQ Index, and transforming of continuous queries into key-values. The Reduce module also should be changed for keeping partial piplelined aggregation for a given time interval.

4 Conclusion

This paper mainly concerns about finding a good approach to using the MapReduce framework for processing multi-dimensional continuous query. we proposed a new idea for transforming overlapped regions of multi-dimensional continuous queries into multiple key values. At the Map stage, input data stream would be partitioned into multiple map tasks which transform input sensor data into corresponding multiple key values. It is easy to group the same key values at the Reduce stage for providing the result of continuous query.

The next step is to verify our approach to generating key values based on multi-dimensional continuous query for both increased continuous query and huge volume of sensor data.

Acknowledgement. This research was supported by the MKE(The Ministry of Knowledge Economy), Korea, under the ITRC(Information Technology Research Center) support program (NIPA-2013-(H0301-13-1012)) supervised by the NIPA(National IT Industry Promotion Agency).

References

1. Dean, J., Ghemawat, S.: Mapreduce: simplified data processing on large clusters. In: OSDI, pp. 137–150 (2004)
2. Hadoop MapReduce, http://hadoop.apache.org/mapreduce/
3. Condie, T., Conway, N., Alvaro, P., Hellerstein, J.M.: MapReduce Online. In: NSDI 2010 (2010)
4. Witkowski, A., Bellamkonda, S., Li, H.G., Liang, V., Sheng, L., Smith, W., Subramanian, S., Terry, J., Yu, T.F.: Continuous Queries in Oracle. In: VLDB, September 23-28 (2007)
5. Arasu, A., Babu, S., Widom, J.: The CQL continuous query language: Semantic foundations and query execution. The VLDB Journal 15(2), 121–142 (2006)
6. Golab, L., Ozsu, M.T.: Processing sliding window multi-joins in continuous queries over data streams. In: Proc. Intl. Conf. on Very Large Data Bases, pp. 500–511 (2003)
7. Babu, S., Widom, J.: Continuous queries over data streams. SIGMOD Record 30, 109–120 (2001)

Measuring Proximity in a Graph of Spatial Data (ZIP Codes)

M. Harist Murdani[1] and Joonho Kwon[2]

[1] Dept. of Computer Engineering, Pusan National University, South Korea
[2] Inst. of Logistics Information Technology, Pusan National University, South Korea
{hariste,jhkwon}@pusan.ac.kr

Abstract. Proximity is used for showing closeness between objects. It is widely used in a social network, marketing, and online businesses. Proximity is also used for spatial analysis such as epidemic spread, extent of trade areas, and socioeconomic planning. Proximity in spatial data is always associated with distance. However, distance itself can be ambiguous when there is an impassable natural landmark barrier between them such as mountain and big lake. Impassable natural landmark barrier is not included in the boundaries of a ZIP Code. Adjacency between ZIP Codes would not happen if they did not have common boundaries. We propose a novel way of measuring proximity in a graph of spatial data based on common boundary points, centroid distance, and shortest path. By considering these three, we bound to get more exact precision answer for proximity and eliminating the possibility of ambiguous distance. We also provide a mathematical model for effectively calculating proximity by using the previously mentioned features.

1 Introduction

Graphs can be used to model various types of dataset in broad categories. Complex relationships such as humans DNA, chemical bonds, social network, spatial data, etc, can be encoded into graphs. Definition of graph usually associated with vertex and edge. The interesting part about graphs is not only the types of connectivity, but also the properties or attributes of the vertex and edge. Nowadays dataset are becoming more complex. This leads to the use of graph as the properties graph, where each vertex and edge is having their own properties.

Proximity measure is a measure of closeness or similarity between two objects. It is not so hard to measure the proximity of two different shapes of objects. However, measuring proximity in a graph of networks is kind of leveled up the challenges. We have to consider for disjointness, distance, and the shortest path between them. In spatial analysis, proximity is used for analyzing epidemiological data, socioeconomic planning, and retailing. Analyzing epidemiological data means mapping of patterns, causes, and effects of health and disease conditions in a predefined regions. Socioeconomic planning is focus on social impact mapping as the result of some sort of economic change. Retailing used to identify the spatial extent of trades area.

B. Hong et al. (Eds.): DASFAA Workshops 2013, LNCS 7827, pp. 79–85, 2013.
© Springer-Verlag Berlin Heidelberg 2013

In a graph of social network, the use of proximity falls to three category: person search[3], link prediction[4,5], and finding clusters or community of entities. Person search is the problem of finding relevant people in a social network. Link prediction is the problem of predicting the new relationships between two objects that are likely related. It is used as a basis for friends recommendation. Finding clusters or community of entities is the problem of clustering the social networks based on the properties of each person such as: schools, address, workplace, etc. Proximity can also be used in online businesses, where we can look for likely suitable person for the purpose of selling some products. Brief comparison about proximity in social network is given in [2].

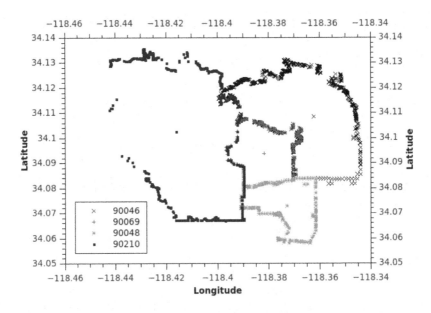

Fig. 1. ZIP Codes Point Map (Boundary and Centroid)

However, proximity measurement in a graph of social network can not be directly applied to a graph of spatial data. Spatial data have latitude and longitude that explains the positions and real distance of a spatial object on earth. As an example of spatial data, we choose the ZIP Code dataset. Fig.1 shows the point map of 4 adjacent ZIP Codes with its boundaries and centroid.

As distance means everything for spatial data, it is closely related and always associated with proximity. Thus proximity computation nave approach in spatial data is the distance between the spatial object. However, distance itself can be ambiguous when there is an impassable natural landmark barrier between them such as mountain and big lake. Impassable natural landmark barrier is not included in the boundaries of a ZIP Code. Adjacency between ZIP Codes would not happen if they did not have common boundaries. Spatial feature leads to the connection between two ZIP Codes that give rise to the concept of common boundary points, centroid distance, and shortest path. Our proximity measurement and mathematical model relies on these three features.

We explained about the spatial data, ZIP Codes, and how to model the ZIP Codes into a graph in Sect.2. In Sect.3, we mentioned the related work that correlated with the basic proximity measurement. We explained in details about our novel way and mathematical model for measuring proximity in a graph of spatial data in Sect.4 We concluded our findings in Sect.5.

2   Graph Spatial Data Modeling

Characteristic of spatial data is the existing of latitude and longitude as part of the data. Spatial Data Types (SDT) is commonly used to model geometry and represent spatial data. SDT provides a fundamental abstraction for modeling the real spatial objects, either single objects or spatially related objects, into any kind of system. One example of spatial data that can be used for measuring proximity is ZIP Codes. ZIP Codes are a system of postal codes used and maintain by the United States Postal Service (USPS). The basic format consists of five decimal numerical digits. Our ZIP Codes dataset are maintained by MapTechnica [7]. In this dataset, ZIP Code is represented by polygon or multi-polygon. Each polygon are made of boundary points. They provide the centroid and the polygon or multi-polygon of each ZIP Code.

Simple graph is represented as $G = (V, E)$, where V is the vertex or node and E is the edge or relationship between nodes. We will use properties graph, each vertex and edge have their own properties, to model the ZIP Codes spatial data into graph-like format. For vertex, there will be two vertex types: *ZIP Codes* and *Boundary Points*. For the edges, there will be also two edge types: *IsAdjacentTo* and *IsLimitedBy*

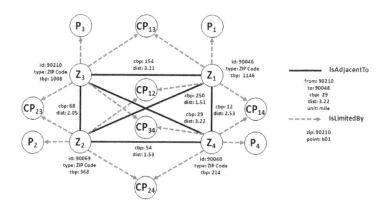

Fig. 2. Graph of ZIP Code

ZIP Codes node are the ZIP Code itself. *Boundary Points* are all pairs of latitude and longitude that build the ZIP Code polygon. *IsAdjacentTo* edge is a relation between two ZIP Code nodes. It is denoted by the existing of common boundary points, same boundary points, between them. *IsLimitedBy* edge is a relation between a ZIP Codes node and a Boundary Points node. Fig.2 shows an example of the ZIP Code graph modeling based on Fig. 1. Z_n is the ZIP Code node. CP_{xy} is the simplified illustration of common boundary point between node Z_x and Z_y. P_n is the simplified illustration of other boundary point. Solid line indicates *IsAdjacentTo* while dashed line indicates *IsLimitedBy* relationship. *tbp* is the total number of boundary points in a ZIP Code node and *cbp* is the total number of common boundary points between two adjacent ZIP Code.

3   Related Work

Proximity measurement in a graph is related to the distance between two nodes. Simplest representation of graph distance is *shortest path*. From the graph theory, the definition of proximity would be the length of the shortest path that connecting two nodes. Using this definition, proximity decays as the distance further apart and disjoint nodes would not be considered because they are unlikely related. However, it is not enough to be used as proximity measurement in a graph of spatial data. Shortest path did not consider the existing of multi-path and common boundary points, which defines the *real relationships*, between two ZIP Codes.

Another measurement for proximity is *node neighborhoods*. The simplest neighborhood-based measure the similarity by means of the neighborhood number. If both have high degree or low degree then their similarity is high rather than if one has high degree while the others have low degree. If s and t have a lot of common neighbors, it means that they are very likely to be closely related each other.

In [6], proximity is being modeled as Effective Conductance (EC) within an electrical circuit, where edges as resistors and edge conductance is proportional to its weights. Higher weight will add more electricity, thus closer proximity. EC has an equal intuitive definition in terms of random walks in graph. EC also has a monotonicity property which states that increasing the conductance or adding new resistor will increase the conductance value between any related two nodes in the network. However it suffers from the existing of degree-1 nodes. Further improvement is in [1]. The authors are improving the effective conductance proximity measurement by using Cycle-Free Effective Conductance (CFEC), which modify the monotonicity of EC for dead end paths, and proximity graphs.

4   Graph Spatial Proximity Measurement

As mentioned in the previous section, proximity measurement for a graph of spatial data is affected by common boundary points, centroid distance, and shortest path. To simplify the example graph of spatial data, from now on we use the ZIP Code and its related terms as explained in section 2.

Borrowing the term of spatial auto-correlation, common boundary points are connecting one spatial area with others. More point means more area for inter-action between them. The neighborhood proximity can be ranked based on the number of common boundary points. As known in social network, if two people have more common friends would means that they are in the same group or living in the same area. Common boundary points indicates the adjacency between to ZIP Codes. For non-adjacent nodes, centroid distance and shortest path are the solution. Calculate the traversal path and centroid distance between two nodes that has the minimum number of nodes and edges weight, which measured using common boundary, altogether.

The precise definition of a graph is as follows. A *graph* is a triple $G = (V, E, \phi)$ where V is a finite set, we called it the vertices of G. E is a finite set, we called it the edges of G, and ϕ is a function with domain E and codomain $P_2(V)$. A *properties graph* is having the same definition of *graph* with the addition of $V\{(VP_1), (VP_2), ..., (VP_n)\}$ where VP is the set of properties owned by vertices V. And also $E\{(EP_1), (EP_2), ..., (EP_n)\}$ where EP is the set of properties owned by edges E.

Definition 1. *For adjacent ZIP Codes, their proximity value can be computed using*

$$AdjProx(\overrightarrow{xy}) = \frac{\sum (p_x \cap p_y)}{\sum p_x} \tag{1}$$

where p_x are the boundary points of x and p_y are the boundary point of y. Using this equation, proximity values between $x \to y$ and $y \to x$ are different. The proximity value will be used as weight (w) of the adjacent edges.

Example 1. Let us use the ZIP Code graph from Fig. 2 to measure the proximity between "90069" (Z_2) and "90046" (Z_1).

$$AdjProx(\overrightarrow{Z_1 Z_2}) = \frac{250}{1146} = 0.22$$

$$AdjProx(\overrightarrow{Z_2 Z_1}) = \frac{250}{368} = 0.68$$

Definition 2. *Proximity value decays as the distance increase by factor of k, where k is the number of node or hop in the shortest path that needs to be tra-versed from $x \to y$. We also consider the real distance between them by compute the source ZIP Code radius against closest adjacent node in the shortest path. Add it as multiplicative factor for measuring proximity in fraction with distance from outer boundary source to target. For non-adjacent ZIP Codes, we compute their proximity value by using*

$$rad_x = \frac{dist_{V_1}}{2} \tag{2}$$

$$SpProx(\overrightarrow{xy}) = \left(\frac{rad_x}{dist_{xy} - rad_x} \times \frac{\sum AdjProx(V_i)}{k} \right) \tag{3}$$

where rad_x is the distance from source node centroid to its outer boundary, $distV_1$ is the distance from source node to the first node in shortest path, $dist_{xy}$ is the distance from source node centroid to target node centroid, and k is the number of node in shortest path.

Example 2. Using shortest path example from Fig. 3, we measure the proximity between "90210" (Z_3) and "91504" (Z_7) as follows:

$$AdjProx(V_1) = AdjProx(\overrightarrow{Z_3Z_1}) = \frac{154}{1008} = 0.15$$

$$AdjProx(V_2) = AdjProx(\overrightarrow{Z_1Z_5}) = \frac{276}{1146} = 0.24$$

$$AdjProx(V_3) = AdjProx(\overrightarrow{Z_5Z_6}) = \frac{35}{721} = 0.05$$

$$AdjProx(V_4) = AdjProx(\overrightarrow{Z_6Z_7}) = \frac{23}{237} = 0.1$$

$$rad_{Z_3} = \frac{distV_1}{2} = \frac{3.11}{2} = 1.55$$

$$SpProx(\overrightarrow{Z_3Z_7}) = \left(\frac{1.55}{10.61 - 1.55} \times \frac{0.15 + 0.24 + 0.05 + 0.1}{4} \right) = 0.023$$

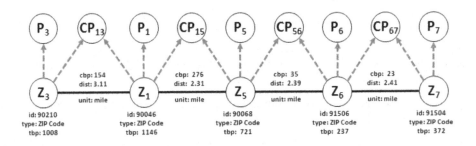

Fig. 3. Example of a ZIP Code Graph Shortest Path

Algorithm 1. Proximity Measurement

Input: Z_s, ZIP Code source; Z_t, ZIP Code target
Output: $prox$, proximity value
1: **if** $IsAdjacentTo(Z_s, Z_t)$ **then**
2:    $prox \leftarrow AdjProx(\overrightarrow{Z_sZ_t})$
3: **else**
4:    $\{V_z\} \leftarrow ShortestPath(Z_s, Z_t)$
5:    $prox \leftarrow SpProx(\overrightarrow{V_z})$
6: **end if**
7: **return** $prox$

---

5   Conclusion

Proximity have many uses based on the graph they are entitled to. In a graph of social network, proximity can be used for person search, link prediction, and finding clusters or community of entities. Also proximity can be applied for marketing and business sale. In spatial analysis, proximity is used for analyzing epidemiological data, socioeconomic planning, and retailing.

However the measurement used in graph of social network can not be applied directly into graph of spatial data. Proximity in spatial data is always associated with distance. However, distance itself can be ambiguous when there is an impassable natural landmark barrier between them such as mountain and big lake. So we proposed a novel way to measure the proximity using common boundary points, centroid distance, and shortest path. We also provide the mathematical model for measuring the proximity using these properties.

To further avoid the distance ambiguity, we can consider the highway road network. The highway road networks are connecting each designated areas which does not have an impassable natural landmark barrier. If it does, it means that there is a man made tunnel or bridge in use. So for our future works, we will add the highway road network into the model and do some experiment to validate or revise our model.

Acknowledgement. This research was supported by the MSIP(Ministry of Science, ICT&Future Planning), Korea, under the ITRC(Information Technology Research Center)) support program (NIPA-2013-(H0301-13-1012)) supervised by the NIPA(National IT Industry Promotion Agency).

References

1. Koren, Y., North, S.C., Volinsky, C.: Measuring and Extracting Proximity Graphs in Networks. In: ACM (2007)
2. Cohen, S., Kimelfeld, B., Koutrika, G.: A Survey on Proximity Measures for Social Networks. In: Ceri, S., Brambilla, M. (eds.) Search Computing III. LNCS, vol. 7538, pp. 191–206. Springer, Heidelberg (2012)
3. Carmel, D., Zwerdling, N., Guy, I., Ofek-Koifman, S., Har'el, N., Ronen, I., Uziel, E., Yogev, S., Chernov, S.: Personalized social search based on the user's social network. In: CIKM, pp. 1227–1236 (2009)
4. Liben-Nowell, D., Kleinberg, J.M.: The link-prediction problem for social networks. JASIST 58(7), 1019–1031 (2007)
5. Zhou, T., Lu, L., Zhang, Y.C.: Predicting missing links via local information. The European Physical Journal 71(4), 623–630 (2009)
6. Faloutsos, C., McCurley, K.S., Tomkins, A.: Fast discovery of connection subgraphs. In: ACM SIGKDD Conference, pp. 118–127 (2004)
7. MapTechnica, www.maptechnica.com

An Efficient Data Access Method Exploiting Quadtrees on MapReduce Frameworks

Hyunho Noh and Jun-Ki Min

School of Computer Science and Engineering,
Korea University of Technology and Education,
Byeongcheon-myeon, Cheonan, Chungnam, Republic of Korea, 330-708
{oksknoh,jkmin}@koreatech.ac.kr

Abstract. Due to the advance of diverse techniques such as social networks and sensor networks, the volume of data to be processed has rapidly increased. After Google proposed the MapReduce framework which processes big data using large clusters of commodity machines, the MapReduce framework is considered as an effective processing paradigm for a massive data set. However, in the view of the performance, a problem of the MapReduce framework is that an efficient access method (i.e., an index) is not supported. Thus, whole data should be retrieved even though a user wants to access a small portion of data. In this paper, we propose an efficient method constructing quadtrees on the MapReduce framework. Our technique reduces the index construction time utilizing a sampling technique to partition a data set. In addition, using the constructed quadtree as well as the MapReduce framework, a subset of data to be retrieved is easily identified and is processed in parallel. Our experimental result demonstrates the efficiency of our proposed algorithm with diverse environments.

Keywords: quadtree, MapReduce, big data.

1 Introduction

Recently, due to the emerging of social networks, advance of sensor technology, proliferation of smart phones, and so on, the volume of data has extremely increased. Since the limited computing power, main memory capability and disk space of a single machine, the performance of traditional algorithm running on a single machine will degrade as the size of data increases.

MapReduce [1] is a distributed parallel processing model and execution environment that proposed by Google for processing massive data sets running on large clusters of commodity machines. Due to its simplicity, flexibility, fault tolerance and scalability, MapReduce becomes to be widely adapted in both commercial and scientific applications using massive data sets, especially multi-dimensional data sets. Therefore, much research has been conducted to migrate traditional algorithms onto the MapReduce framework such as Wavelet Histogram [2], Join operator [3,4] spatial data processing [5] and so on.

B. Hong et al. (Eds.): DASFAA Workshops 2013, LNCS 7827, pp. 86–100, 2013.

The commonly argued issue of above work is that there is no functionality provided for building and accessing such as spatial indexes in the MapReduce framework. Thus, even though a user would like to access a small portion of data, whole data should be retrieved in the MapReduce framework.

In this paper, to support efficient access to a large sized d-dimensional data set, we propose a quadtree construction algorithm on the MapReduce framework and an access method using the constructed quadtree which is also running on the MapReduce framework.

Since a quadtree [6] is conceptually simple and easy to maintain, we choose a quadtree as a data access method. In addition, for constructing an index of a large sized data set in a single machine, much time is required and it may be hard to keep intermediate index structures. Therefore, to solve above problems, we utilize the MapReduce framework for constructing a quadtree.

Of particular, after constructing a quadtree, since we can rapidly reduce the search space using the constructed quadtree and, in nature, apply the MapReduce framework to process data in parallel, we can improve the search performance.

The remainder of the paper is organized as follows. In Section 2, we present the background of our work. Section 3 introduces the quadtree construction on the MapReduce framework. Section 4 describes the access scheme using the constructed quadtree. Section 5 presents an empirical evaluation. Section 6 discusses related work and Section 7 summarizes the paper.

2   Preliminary

2.1   QuadTrees

For databases, several index structures such as B+-tree [7] and R*-tree [8] have been proposed. As one of the prominent index structures for multi-dimensional data, a *quadtree* [6] is used in diverse areas such as spatial data management and image processing.

Given a d-dimensional point data set D, a quadtree (2^d tree where $d \geq 2$) is constructed by recursive partitioning until every leaf node has less than c points where c is called the *capacity*. Every node in a quadtree represents a d-dimensional space. By partitioning, a d-dimensional space associated with a node n is divided into equi-sized 2^d subspaces each of which is represented by a n's child node. Thus, every intermediate node of a quadtree has 2^d child nodes and every leaf node n_l has a subset of points which are located in the region presented by n_l. Figure 1 shows a quadtree for a 2-dimensional data set with $c = 2$.

2.2   MapReduce

MapReduce [1] is a distributed and parallel processing model and execution environment in shared-nothing clusters of machines. Hadoop [9] is implemented in

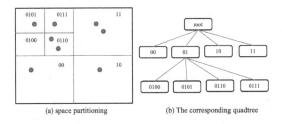

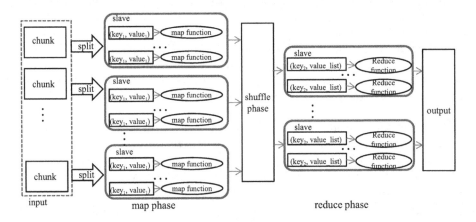

Fig. 1. An Example of Quadtrees

Fig. 2. The Data Flow in MapReduce

the OpenSource commnuity as the MapReduce framework. In Hadoop, using the hadoop distributed file system (HDFS), a large sized file is initially partitioned into several fragments, called *chunk*, and stored in several machines redundantly for reliability. The size of a chunk is typically 64 MByte.

In MapReduce, a program consists of a map function and a reduce function which are user-defined functions. Basically, the MapReduce framework consists of one *job tracker* and several *task trackers*. Each task tracker is running on a commodity machine, called *slave*. A slave processes data using a map function and a reduce function each of which is invoked by the task tracker. The job tracker running on a single machine, called *master*, takes responsibility for error detection, load balancing and so on.

The data processing in MapReduce is composed of three phases: *map phase*, *shuffle phases*, and *reduce phase*. Figure 2 illustrates the data flow in the MapReduce framework.

Map Phase: A task tracker in the map phase is called a *mapper*. A map function invoked by a mapper takes a key-value pair $(key_1, value_1)$ as input, executes some computation, and may output a set of intermediate key-value pairs $(key_2, value_2)$. In addition, by applying a combine function, additional computation to intermediate results is executed.

Shuffle Phase: In this phase, intermediate key-value pairs are grouped with respect to the key k_2. Thus, a reduce function to be executed in the next phase can obtain a list of values having the same key. Therefore, through this phase, each work is assigned the data lists as $(key_2, value\_list)$.

Reduce Phase: A task tracker in the reduce phase is called a *reducer*. The reducer in a slave invokes a reduce function with a key and a value list. Each reduce function executes computation for the value list and emits a key-value pair $(key_3, value_3)$ as a final result.

3   QuadTree Construction on MapReduce

For a large volumed data set, if a quadtree in constructed on a single machine, much time is required and, perhaps, intermediate results generated during the quadtree construction cannot be kept in a single machine.

In this section, we present the quadtree construction method for a large sized multi-dimensional data set exploiting the MapReduce framework. Strictly speaking, we explain how the quadtree Q_c^D for a data set D with the capacity c is constructed using MapReduce.

3.1   Naive Method

Since MapReduce is running on the shared-nothing cluster of machines, each machine cannot share the information of the other machine's processing status. Thus, the data set should be separated into several subgroups each of which is allocated to a slave.

In this section, we explain a naive method to construct a quadtree on the MapReduce framework. In the naive method, the d-dimensional data space is partitioned into the predefined d-dimensional subspaces. we call this naive method QMR (Quadtree with MapReduce). QMR consists of two steps. In the first step, each point p is distributed with respect to the grid containing p by the map function and the quadtree for each grid is constructed by the reduce function. In the second phase, the quadtrees are consolidated into a final quadtree in a single machine.

In QMR, under the assumption that the points are uniformly distributed in the data space, the data space is partitioned into several equi-sized subspaces. Thus, QMR will show the reasonable performance when the points are uniformly distributed since the number of points in each partition is similar to that of the other partitions. The data partitioning strategy using predefined grids is applied in much work [10,11].

Figure 3 shows the partitioning of 2-dimensional data space with the predefined grids in QMR. As shown in Figure 3, since the data space is blindly partitioned without the information of the data distribution, the numbers of points in partitions could be quite different. In this case, a slave assigned the partition containing a lot of points consumes much time to construct the quadtree for the partition, and hence, overall performance of QMR is degraded. In addition,

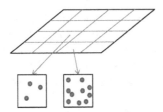

Fig. 3. Predefined Data Partitioning

when the number of partitions is less than the number of slaves, some machines can not participate in the quadtree construction. For instance, in [11], the d-dimensional data space is partitioned into 2^d subspaces. Thus, when $d = 2$, the number of partitions is 4. In this situation, the MapReduce framework is not fully utilized when the number of machines in a cluster is greater than 4.

Therefore, in the next section, we propose an efficient quadtree construction method on the MapReduce framework, which utilizes the sampling techniques in order to identify the data distribution approximately.

3.2 QuadTree Construction with Sampling

As mentioned in Section 3.1, if the data space is divided without the information of the data distribution, the performance of the system will be degraded.

Given a d-dimensional data set D whose size is $|D|$, in order to construct efficiently the quadtree Q_c^D with the capacity c on the MapReduce framework, we should make an effective partitioning method. The partitioning method should satisfy the following requirements.

(1) The number of points belonging to a partition should be similar to those of the other partitions in order not to pose the large overhead to a certain slave.
(2) In order to maximize the system utilization, the number of partitions should be greater than or equal to the number of slaves t.
(3) The quadtrees for the partitions should be easily consolidated into the final quadtree.

To partition the data space, we use another quadtree, called the base quadtree (See Definition 2). As shown in Figure 1, the quadtree partitions the data space where each leaf node has the points less than the capacity. Thus, the numbers of points of the partitions tend to be similar to each other. Furthermore, when we construct the quadtree for each partition, since the partition is associated with a leaf node of the base quadtree, the requirement (3) can be easily satisfied.

We call the our proposed quadtree construction method $SQMR$ (Sample based Quadtree with MapReduce). The pseudo code of SQMR is presented in Figure 5. In order to remove the confusion, we define the some notations of quadtrees.

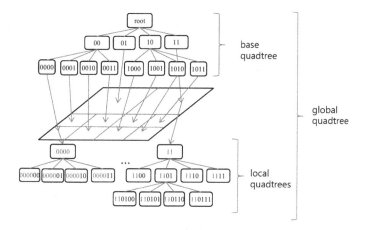

Fig. 4. The Hierarchical Structure of the Global Quadtree

Definition 1. *Given d-dimensional point set D whose size is $|D|$, the global quadtree is the quadtree Q_c^D constructed with D where the capacity is c.*

Definition 2. *Given d-dimensional point set D, the base quadtree is the quadtree $Q_{|D|/t}^D$ constructed with D where the capacity is $|D|/t$.*

Definition 3. *Given a leaf node n of the base quadtree and the d-dimensional point set D, the local quadtree is the quadtree Q_c^P constructed with a set of points $P \subset D$ which are located in the region represented by n where the capacity c.*

Figure 4 illustrates the hierarchical structure of the global quadtree which is composed of the base quadtree and a set of local quadtrees.

By Definition 2, the capacity of the base quadtree is $|D|/t$. Thus, although the base quadtree does not strictly satisfy the requirement (1), every leaf node of the base quadtree has at most $|D|/t$ points. Additionally, since the sum of the numbers of points in all leaf nodes should be $|D|$, the number of leaf nodes is greater than or equal to the number of machines t. Therefore, the requirement (2) is satisfied.

To construct the base quadtree, all points should be retrieved. For a large sized data set, the base quadtree construction also requires much time. Therefore, we propose the data partition method using the random sampling.

SQMR is composed of three steps: *preprocessing step, local quadtree construction step*, and *consolidation step*.

Preprocessing Step: Since building the base quadtree with large data in very expensive disk I/Os, we generate a sample set S from the data D using the reservoir sampling [12]. An approximate base quadtree is constructed using S in a single machine.

```
Function SQMR( D, c, d, t, f )
D: a data set, c: capacity, d: the dimension
t: the number of machines, f: sampling ratio
begin
        //Preprocessing Step
1.  S = ReservoirSampling(D, f );
2.  Q^S_{|S|/t} = QuadTree( S, |S|/t );
        //Local QuadTree Construction Step
3.  Broadcast Q^S_s and c;
4.  LocalTrees = RunMapReduce(LOCALQ.map, LOCALQ.reduce);
        //Consolidation Step
5.  QuadTree = GLOBALQ(Q^S_s, LocalTrees );
6.  return QuadTree;
end
```

Fig. 5. The algorithm of SQMR

Local Quadtree Construction Step: In this step, the points in D is split based on the regions divided by the approximate base quadtree and the local quadtree for each partition is computed independently in parallel.

Consolidation Step: Finally, we construct the global quadtree using the local quadtrees for the partitions. Since each partition associates with each leaf node of the approximate base quadtree, by replacing the leaf node of the approximate base quadtree to the proper local quadtree's root node, the global quadtree is easily constructed.

Now, we explain the above three steps in detail.

Preprocessing Step. Since the base quadtree construction consumes much time, we construct the *approximate base quadtree* using the set of sample points S.

In our initial work, we try to perform the data sampling on the MapReduce framework. In other words, the data sampling is conducted in the map phase and, in the reduce phase, the approximate base quadtree is constructed by a single reduce function.

However, in the map phase, each map function is invoked with each point p as an input. And a map function emits p or not with respect to the sampling ratio f. Therefore, although the data set is retrieved in parallel on the MapReduce framework, the benefit of the sampling is marginal since every point should be retrieved at least once.

For such a reason, in our work, the data sampling is performed in a single machine (line 1 of SQMR in Figure 5). To sample points, we adopt the reservoir sampling [12] which is devised as a sampling technique for stream data. The reservoir sampling is a family of randomized algorithms for randomly choosing $|S|$ samples from a data set D.

In the reservoir sampling, initially, the first $|S|$ points are selected as the sample points. Then, the rest of points in D is processed. Instead of flipping a coin for each point $p \in D$ to add p as a sample, the number of skipped points before the next point to be added to S is determined with respect to the sampling ratio $f(=|S|/|D|)$. When a new point q is selected as a sample point, a randomly chosen point p in S is evicted and q is inserted into S. Since some points in D is skipped without retrieval, the reservoir sampling is adequate for a large sized data set.

We construct the approximate base quadtree $Q^S_{|S|/t}$ where the capacity is $|S|/t$ (Line 2 of SQMR). Thus, like the base quadtree, the number of $Q^S_{|S|/t}$'s leaf nodes is greater than or equal to the number of machines t and the integration of local quadtrees into the global query tree is simple. Therefore, the requirement (2) and (3) are satisfied.

For the sufficiently large number of sample points, the sample point set S reflects the data distribution of D. The variance is a measure of how far a set of numbers is spread out.

In the statistical view, when $|S| > 30$, by the central limit theorem, the expected sample mean is population mean and the expected sample variance is equal to the population variance divided by $|S|$-1 [13]. Thus, the variance of sample points in each dimension is an unbiased estimator for the variance of D in each dimension. In addition, the $d \times d$ covariance matrix of S is also an unbiased estimator for the covariance matrix of D [14]. Thus, the shape of the approximate base quadtree will be similar to that of the base quadtree.

Local QaudTree Construction Step. The local quadtree is constructed in this step. The pseudo code for the local quadtree construction is presented in Figure 6.

The approximate base quadtree $Q^S_{|S|/t}$ and the capacity c are first broadcast to all slaves (line 3 of SQMR in Figure 5). The task tracker invokes the map function with a point $p \in D$. Each map function find the leaf node n of $Q^S_{|S|/t}$ whose corresponding region contains p (line 2 of LOCALQ.map in Figure 6). And the map function emits the key-value pair (n, p) (line 3 in LOCAL.map).

As mentioned in Section 2, during the shuffle phase, the key-value pairs generated by all map function are sorted and grouped by key. The reduce function is called with each key key and the corresponding point list P. Then the reduce function constructs the local quadtree Q^P_c with the capacity c by invoking $QuadTree(P, c)$ and outputs the local quadtree Q^P_c and key (i.e, node id of the approximate base quadtree) (lines 1-3 of LOCALQ.reduce in Figure 6).

In the legacy systems, a page is the unit of disk I/O. Thus, in the traditional tree structured indexes, a page is used for a node. The typical size of a page is 16KByte or 64KByte. However, HDFS manages a file with large sized units called chunks since, if data are split into a lot of physical/logical units, it is hard to maintain the corresponding meta data. As mentioned earlier, the typical size of a chunk is 64MByte. Therefore, in our work, we set the capacity c of the local quadtree (as well as the global quadtree) 64Mbyte.

In addition, when a local qudtree is constructed, each set of points belonging to each leaf node of a local quadtree is stored into a separated file and the leaf node only records the file name. Thus, we can obtain the data clustering effect.

Function LOCALQ.map(key, p)
key: null, p: a point
begin
1. $Q^S_{|S|/t}$ = LoadQuadTree();
2. n = FindLeafNode(p, $Q^S_{|S|/t}$)
3. emit(n, p)
end

Function LOCALQ.reduce(key, P)
key: a node id, P: a list of point
begin
1. c = LoadCapacity();
2. Q^P_c = QuadTree(P, c);
3. output(key, Q^P_c);
end

Fig. 6. The Algorithm for Local Quadtree Construction

Function GLOBALQ($Q^S_{|S|/t}$, $\{(key, Q^P_c)\}$)
$Q^S_{|S|/t}$: the approximate base quadtree
$\{(key, Q^P_c)\}$: a set of local quadtrees with corresponding keys
begin
1. **for each** $(key, Q^P_c) \in \{(key, Q^P_c)\}$ **do**
2. n = FindNode(key, $Q^S_{|S|/t}$);
3. n_p = n.parentNode;
4. $n_p.n$ = Q^P_c.root;
end

Fig. 7. The Algorithm for the Consolidation

When a query is submitted, by the quadtree traversal, we can find the proper files which contain the query result.

Consolidation Step. In this step, the local quadtrees generated in the previous step are integrated into the global quadtree. The pseudo code for this step is presented in Figure 7.

The function GLOBALQ takes the approximate base quadtree and the set of local quadtrees each of which has a key. Since the key is the leaf node id of the approximate base quadtree, the corresponding leaf node of the approximate base quadtree is easily identified (line 2 of GLOBALQ in Figure 7. Then the leaf node is replaced by the root node of the local quadtree (lines 2-3 of GLOBALQ). By applying this procedure to all local quadtrees, the global quadtree is constructed.

After the global quadtree is constructed, the maintenance of the global quadtree by insertion or deletion of a point is simple. When a new point is inserted into a leaf node, if the number of points in the leaf node exceeds the capacity, the node is split into 2^d chid nodes. When a point in a leaf node is removed, if the sum of numbers of points belonging to itself and its siblings is less than the capacity, the node and the siblings are merged into its parent node. Since it is trivial, we omit the details.

4   Data Access via Quadtree on MapReduce

In this section, we explain how to utilize the constructed quadtree to access the large sized data.

To retrieve points in the data set D, a point query or a range query is submitted. To process the submitted query using the MapReduce framework, the reduce phase is not require [15] since each map function with a point p as an input generates p if p satisfy the query condition. However, when the MapReduce framework is used only, whole points should be retrieved in the map phase to process the submitted query. In contrast, using the quadtree, we can improve the performance since the search space is significantly reduced.

For a point query, we can easily identify the leaf node whose corresponding region contains the query point. Since the leaf node contains the name of the file which contains all points belonging to the leaf node, we can efficiently obtain the query result by retrieval of the file (i.e, a subset of D).

For a range query, the lead nodes overlapped with the query range are identified by traversal of the quadtree. Therefore, we can obtain the set of files. As illustrated in the wordcount program, a well known MapReduce example [1], MapReduce can read records from multiple files. Therefore, using the MapReduce framework, we can retrieve the points stored in the several files and generate the query results during the map phase in parallel.

5   Experiment

In this section, we evaluate and compare the performance of our proposed quadtree construction method with the other techniques

5.1   Experimental Environment

To perform the experiment, we set up a cluster of 31 commodity PCs. One of PCs acts as the master and 30 PCs act as slaves. The master has 3.1 GHz Intel Xeon E3-1220 CPU, 16 GByte memory and 500GByte hard-disk. Each slave has 3.2 GHz Intel Core i5 CPU, 4GByte memory and 1TByte hard-disk. All machines are connected through a 1Gbps Ethernet switch. Every node is running on Linux (Ubuntu 10.04 Lucid). We implements all algorithms using JDK 1.6. We use Hadoop 2.0.0 for the MapReduce framework implementation obtained from [9].

To compare the our proposed method SQMR, we implement two algorithms: QSM (Quadtree with a Single Machine) and QMR(Quadtree with MapReduce). To show the efficiency of MapReduce, we use QSM which constructs the quadtree in a single machine. QSM is running on the most efficient machine (i.e., master).

In addition, to demonstrate superiority of our method utilizing the sampling, we use QMR. As presented in Section 3.1, QMR partitions the data space into equi-sized subspaces without considering the data distribution. Since there are 30 slaves in the cluster, we set the number of partitions 32 in order to utilize all slaves.

In our experiment, we use the synthetic data set since it is easy to make diverse environments. To make diverse environments, we use some parameters, as summarized in Table 1.

The domain of each dimension is [0.0, 1000.0]. We generate two types of synthetic data sets: *uniform* and *skewed*. In uniform, the points are randomly distributed in the data space. To make the skewed data set, we used the normal distribution N(300.0, 50). We also vary the number of points from 1.0×10^8 to 5.0×10^8. The sizes for a set of 1.0×10^8 2-dimensional points, a set of 1.0×10^8 4-dimensional points, and a set of 5.0×10^8 8-dimensional points are 1.68GByte, 3.17GByte and 30.73GByte respectively. As presented in Section 3.2, we set the capacity (c) 64Mbyte. For SQMR, we choose 10,000 points as sample points.

Table 1. Parameters

Parameter	Default value	Comments & Range
$dist$	skewed	data distribution (uniform, skewed)
$|D|$	1.0	the number of points $\div 10^8$ (1.0, 2.5. 5.0)
t	20	the number of slaves (10,20,30)
d	4	dimensions $(2 \sim 8)$
c	64MByte	the capacity (64MByte)
$|S|$	10,000	then number of sample points (10,000)

5.2   Experimental Result

We measure the performance of each method five times and report the average execution time in this section.

Data distribution: Figure 8(a) illustrates the quadtree construction time with respect to the data distribution. Of particular, the quadtree construction of the skewed data set takes much more time compared to that of the uniform data set since the node partitioning occurs severely in the skewed data set.

In QMR, the data space is partitioned into the equi-sized subspaces. In contrast, SQMR has the sampling overhead to build the approximate base quadtree. The Preprocessing step of SQMR takes about 34 seconds on the set of 1.0×10^8 4-dimensional points and takes about 187 seconds on the set of 5.0×10^8 8-dimensional points to perform sampling and build the approximate base quadtree.

Since the preprocessing overhead is quite small within the overall quadtree construction time, the performance of SQMR is slightly better than that of QMR in the uniform data set in spite of this overhead. Additionally, for the skewed data set, SQMR shows the better performance compared to QMR. This result indicates that SQMR identifies the data distribution and partitions the data space efficiently during the Preprocessing step.

QSM shows the worst performance compared to the other methods using MapReduce, although QSM is conducted on the most powerful machine (master). Thus, we omit the performance of QSM in the following experimental results.

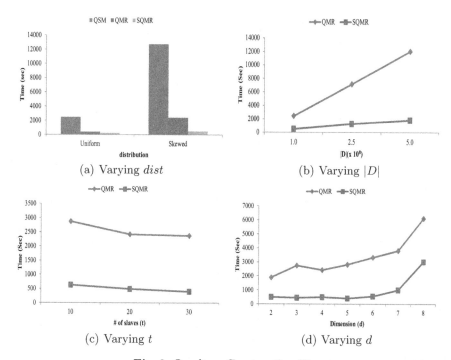

Fig. 8. Quadtree Construction Time

Varying the Number of Points ($|D|$)**:** Figure 8(b) shows the performance with varying the number of points in the skewed data set.

SQMR shows the better performance compared to QMR over all cases. Of particular, the performance gap between SQMR and QMR increases as the number of points increases. Since QMR partitions the data space blindly, the numbers of points in partitions tend to be quite different as $|D|$ increases. In contrast, SQMR partitions the data space such that every partition has the similar number of points. This result indicates the SQMR constructs the quadtree efficiently for the large sized data set.

Varying the Number of Slaves (t)**:** Figure 8(c) shows the performance with varying the number of slaves. As shown in Figure 8(c), the performance of SQMR is improved as the number of slaves increases. Although the performance of QMR is also improved as the number of slaves increases, SQMR shows about 9 times better performance compared to QMR.

Varying the Number of Dimensions (d)**:** Figure 8(d) illustrates the performance varying the number of dimensions. SQMR is also superior to QMR in this experiment. As the number of dimensions increases, the data size increases. Thus the execution times of QMR and SQMR increase as d increases. However, since QMR partitions the data space into the predefined number of subspaces regardless the number of dimensions, the performance of QMR tends to be dramatically degraded as the number of dimensions increases compared to that of SQMR.

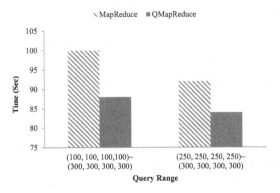

Fig. 9. The Range Query Performance

Data Access. In this experiment, we show the performance of the data access using the quadtree, called *QMapReduce*, compared with the data access without the quadtree, called *MapReduce*.

Since, for a point query, the points belonging to the leaf node whose corresponding region contains the query point are only needed to be retrieved in QMapReduce. Thus QMapReduce is definitely more efficient than MapReduce which retrieves whole points to process the point query. Thus, we show the performance of the range query in this experiment.

Figure 9 show the performance of the range query processing. As the query range increases, the performance of MapReduce and that of QMapReduce slightly increase since the number of query result increases and hence the write overhead in HDFS increases. As shown in Figure 9, QMapReduce is more efficient that MapReduce since QMapReduce reduces the search space efficiently using the quadtree and generates the query results in parallel.

6   Related Work

Recently, many techniques using the MapReduce framework have been proposed. In [16], the raster image sorting technique using the road information in satellite images was proposed. In [17] k-nearest neighbor (k-NN) and reverse nearest neighbor problems are solved using MapReduce based on the Voronoi partitioning. In [3], an efficient k-NN join technique using MapReduce was proposed. In [18], all nearest neighbor query was processed on the MapReduce framework. Of particular, like our work, the data space is divided into several tiles for load balancing and the tiles are integrated according to Z-order. However, if this technique is applied to the quadtree construction, it is hard to consolidate the local quadtrees into the global quadtree. In [19], R-tree was constructed using MapReduce. All object are sorted with respect to a space filling curve such as Z-order and partitioned into the subsets. For each subset, the local R-tree is constructed and the local R-trees are combined into a final R-tree. However, in this work, the maintenance of the constructed R-tree is not mentioned and the access method using the R-tree on the MapReduce framework is not introduced. [20] is the most related to our work. To perform the earthquake simulation, the quadtree for 3-dimensional points is constructed on the MapReduce framework. In [20], the quadtree is constructed in button-up fashion. In the first MapReduce job, each point is transformed to a node and the nodes are merged into partial quadtrees. In next MapReduce job, the partial quadtress are merged. This job is iteratively conducted until the single quadtree is generated. Therefore, since several MapReduce jobs are required to construct the final quadtree, overall performance is degraded.

7   Conclusion

In this paper, we propose an efficient data access method using the quadtree on the MapReduce framework. To partition the data set, we build the approximate base quadtree which is constructed using the sample points. To choose a set of sample points efficiently, we adopt the reservoir sampling. For each partition, we construct the local quadtree on the MapReduce framework. We present that the local quadtrees are easily integrated into the global quadtree since each partition is associated with each leaf node of the approximate base quadtree. In addition, we explain the constructed quadtree is utilized to process a point query as well as a range query on the MapReduce framework. In our experiment, we show the efficiency of the proposed quadtree construction method as well as the efficiency of the data access using the constructed quadtree.

Acknowledgements. This research was supported by Basic Science Research Program through the National Research Foundation of Korea(NRF) funded by the Ministry of Education, Science and Technology (2012R1A1B3003060).

References

1. Dean, J., Ghemawat, S.: Mapreduce: Simplified data processing on large clusters. Communication of the ACM 51(1), 107–113 (2008)
2. Jestes, J., Yi, K., Li, F.: Building wavelet histograms on large data in mapreduce. In: Proceedings of VLDB, pp. 109–120 (2012)
3. Lu, W., Shen, Y., Chen, S., Ooi, B.C.: Efficient processing of k nearest neighbor joins using mapreduce. In: Proceedings of VLDB, pp. 1016–1027 (2012)
4. Zhang, X., Chen, L., Wang, M.: Efficient multi-way theta-join processing using mapreduce. In: Proceedings of VLDB, pp. 1184–1195 (2012)
5. Wang, Y., Wang, S.: Research and implementation on spatial data storage and operation based on hadoop platform. In: Proceedings of International Conference on Geoscience and Remote Sensing, pp. 275–278 (2010)
6. Finkel, R., Bentley, J.: Quad trees a data structure for retrieval on composite keys. Acta Informatica 4(1), 1–9 (1974)
7. Comer, D.: The ubiquitous b-tree. ACM Comput. Surv. 11(2), 121–137 (1979)
8. Beckmann, N., Kriegel, H.P., Schneider, R., Seeger, B.: The r*-tree: An efficient and robust access method for points and rectangles. In: Proceedings of ACM SIGMOD, pp. 322–331 (1990)

9. Apache: Apache hadoop (2010), `http://hadoop.apache.org`
10. Xia, C., Lu, H., Ooi, B.C., Hu, J.: Gorder: An efficient method for knn join processing. In: Proceedings of VLDB, pp. 756–767 (2004)
11. Zhang, B., Zhou, S., Guan, J.: Adapting skyline computation to the mapreduce framework: algorithms and experiments. In: Xu, J., Yu, G., Zhou, S., Unland, R. (eds.) DASFAA Workshops 2011. LNCS, vol. 6637, pp. 403–414. Springer, Heidelberg (2011)
12. Vitter, J.S.: Random sampling with a reservoir. ACM Transactions on Mathematical Software 11(1), 37–57 (1985)
13. Devore, J.L.: Probability and statistics for engineering and the science, 4th edn. Duxbury Press (1995)
14. Vershynin, R.: How close is the sample covariance matrix to the actual covariance matrix? Journal of Theoretical Probability 25(3), 655–686 (2012)
15. Zhang, S., Han, J., Liu, Z., Wang, K., Feng, S.: Spatial queries evaluation with mapreduce. In: Proceedings of GCC, pp. 287–292 (2009)
16. Wu, X., Carceroni, R., Fang, H., Zelinka, S., Kirmse, A.: Automatic alignment of large-scale aerial rasters to road-maps. In: Proceedings of ACM GIS, pp. 17:1–17:8 (2007)
17. Akdogan, A., Demiryurek, U., Banaei-Kashani, F., Shahabi, C.: Voronoi-based geospatial query processing with mapreduce. In: Proceedings of IEEE CloudCom, pp. 9–6 (2010)
18. Wang, K., Han, J., Tu, B., Dai, J., Zhou, W., Song, X.: Accelerating spatial data processing with mapreduce. In: Proceedings of IEEE ICPADS, pp. 229–236 (2010)
19. Cary, A., Sun, Z., Hristidis, V., Rishe, N.: Experiences on processing spatial data with mapreduce. In: Winslett, M. (ed.) SSDBM 2009. LNCS, vol. 5566, pp. 302–319. Springer, Heidelberg (2009)
20. Schlosser, S.W., Ryan, M.P., Taborda, R., López, J., O'Hallaron, D.R., Bielak, J.: Materialized community ground models for large-scale earthquake simulation. In: Proceedings of ACM/IEEE Conference on Supercomputing (2008)

Optimized Adaptive Hybrid Indexing for In-memory Column Stores

Zhongbin Xue[1,2,*], Xiongpai Qin[1,2], Xuan Zhou[1,2],
Shan Wang[1,2], and Anxuan Yu[1,2]

[1] DEKE Lab, Ren-min University of China, Beijing 100872, China
[2] School of Information, Ren-min University of China, Beijing 100872, China
zbxue@rurc.edu.cn

Abstract. Modern applications and databases using dynamic storage environment are characterized by two challenging features: (a) little idle system time to devote in physical design; (b) little priori knowledge about the query and data workload. Traditional approaches to index building and maintenance do not work well in such dynamic environment; while adaptive indexing can be a remedy. An adaptive index is a partially created index. Refinement of the index is conducted during query execution. Database cracking and adaptive merging are two techniques for adaptive indexing. The former is advantageous at initialization, while the latter can converge to its optimal structure with a much faster speed. In this paper, we propose a hybrid approach by combining cracking and adaptive merging. We designed a cost model to measure the cost of data partition operations. Based on the model, we provide an algorithm to refine adaptive index. Experiments show that our hybrid approach can achieve appropriate performance tradeoff between database cracking and adaptive merging.

Keywords: adaptive merging, database cracking, adaptive indexing, hybrid algorithm.

1 Introduction

The deployment of modern applications and databases needs to be fast. This poses two challenges: first, little time can be devoted in physical design; second, little priori workload knowledge can be used, considering that query and data workload can change constantly. Thus traditional approaches to index building and maintenance do not work well in such scenarios. Instead adaptive index is required. The main advantage of adaptive index lied in that it is self-organized and able to optimize itself to the changing query workload dynamically. Given a key range, if it is queried more frequently, its corresponding index structure will be more optimized. If a key range is not queried at all, it will not be optimized.

---

[*] Corresponding author.

B. Hong et al. (Eds.): DASFAA Workshops 2013, LNCS 7827, pp. 101–111, 2013.

Two distinct approaches to adaptive indexing, i.e., database cracking [5,6,7,10]and adaptive merging[3,4], have been proposed recently. Both approaches are implemented in a modern column-store database system [4,5,8]. Experiments showed that adaptive merging usually incurs relatively high initialization cost but converges rapidly, while database cracking enjoys a low initialization cost but converges slowly. A series of hybrid algorithm [8] have been proposed to combine the advantage of the two algorithms.

In this paper, we propose a new hybrid adaptive index to combine the advantages of database cracking and adaptive merging. Our approach is intended to meet both the design goal of database cracking minimizing initial per query overhead and the design goal of adaptive merging in exploiting the concept of runs and merges to converge quickly. It utilizes a cost model to determine the best operation to refine the index such that the best utility is achieved. Experiments show that our hybrid approach can achieve appropriate performance tradeoff between database cracking and adaptive merging.

The rest of the paper is organized as follows. Section 2 introduces and analyzes previous approaches of adaptive indexing. Section 3 then presents our cost model and a hybrid crack crack/sort algorithm of the proposed adaptive index and Section 4 provides a detailed experimental analysis. Finally, Section 5 concludes the paper.

2   Background and Prior Work

Prior Approaches. Most previous approaches to runtime index tuning [11, 12, 13, 14] are non-adaptive, which index tuning and query processing operations are distinct from each other. These approaches decide which indexes to create or drop by monitoring the running workload. When performing index tuning or index creation, both actions impact the database workload. Some research on the query process shows that some data items are more heavily queried than others. So partial indexes[15, 16] were proposed.

Database Cracking. Database cracking combines the features of automatic index selection and partial indexes, and introduce the notation of continuous, incremental, partial and on demand adaptive indexing. Thereby, indexes are incrementally built and refined during query processing. When the first query comes, a new cracker index is initialized. As more queries come, the cracker index is refined by range partitioning until sequentially searching a partition is faster than binary searching.

In a cracker index, keys are partitioned into disjoint key ranges, while the keys remain unsorted within each partition. For each range query, it analyzes the cracker index first, then scans the key ranges that fall entirely within the query range, and uses the two end points of the query range to further partition the keys into two key ranges. Each partitioning step creates two new sub-partitions using logic similar to partitioning in quicksort[17]in most cases. But if both end points fall into the same key range the range is partitioned into 3sub-partitions.

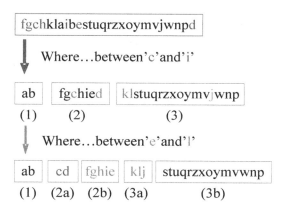

Fig. 1. Database Cracking

The example in Figure 1 shows data is unsorted in the initial array. As the first query on the range "c-i" comes, the array is split into three partitions: (1) keys before 'c'; (2) keys between 'c' and 'i'; and (3) keys after 'i'. Then a new query asking for range "e – l" comes. The first partition can be ignored, as it does not overlap with the query range. Partitions (2) and (3)are further cracked on keys 'e' and 'l', respectively. Subsequent queries continue to partition these key ranges until the structures are optimized for the workload.

In [6], the authors integrate updates into data structure of cracking. Selection on multi-column is considered in [7]. In addition, [7] supports adaptive space management via partial cracking.

Adaptive Merging. Database cracking functions like an incremental quicksort, each query as a quicksort step, resulting in at most one or two partitioning steps. Adaptive merging functions as an incremental merge sort, with one merge step applied to all the key ranges in a query's result. Under adaptive merging, the first query sorts all the initial partitions and finds the relevant data from the initial partition, using binary search. The data fetched from the initial partitions will be merged into the final partition.

As shown in Figure 2, for instance, as the first query (with range boundaries 'c' and 'i') arrives, the data from each equally-sized partition is loaded in memory and sorted. Then it retrieves relevant values (via index lookup because each partition is sorted) and merges them into a final partition. When a subsequent query arrives, it queries from the final partition. If the query is not fully matched, the initial partitions need to be queried again. Results from a second query on range "e – l" from the initial partitions are then merged into the final partition. Subsequent queries continue to merge results from the runs until the final partition is fully optimized for the workload.

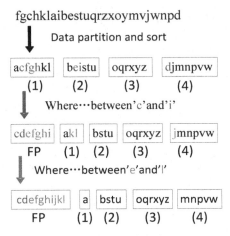

Fig. 2. AdaptiveMerging

Hybrid Approaches. Based on database cracking and adaptive merging, a series of hybrid approaches have been proposed. The hybrid approaches use the query mode in Figure 3.

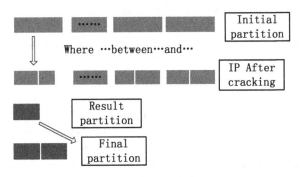

Fig. 3. Query Mode

The hybrid approaches are named by the techniques that used both in the initial and the final partitions. They consider three different techniques of physical reordering tuples in a partition: sorting, cracking and radix clustering. As a result, there are nine potential hybrid algorithms.

Table 1. hybridalgorithm

		Sort	HSS	HSR	HSC
Initial partition		Radix	HRS	HRR	HRC
		Crack	HCS	HCR	HCC
			Sort	Radix	Crack
			Final Partition		

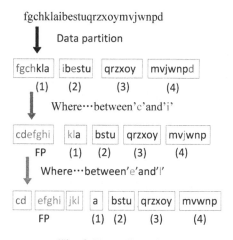

Fig. 4. Hyper Crack Sort

Through extensive experiment[8],researchers found that the Hyper crack crack(HCC) and Hyper crack sort(HCS)are better among the nine hybrid algorithms.

Figure 4 illustrate the process of HCS. Data is loaded and partitioned into different part, unsorted. When the first query arrives, it asks for the range between 'c' and 'i'. In the initial partitions, it uses cracking to find the results. The results "fgchied" are queried from the initial partitions. Then the results are sorted and merged into final partition. So the data in the final partitions is "cdefghi". For the subsequent query, it uses binary search in the final partition to find the result. If the result is not full match then query in the initial partitions and merge the result into final partition, sorted. The different between HCC and HCS is that in HCC data merge to final partition is not sorted.

3   The Cost Model

Based on the experiment results in [8], we obtain a comparison between the HCC and HCS algorithms.

Table 2. Comparsonof HCC and HCS

	The first query response time	Data converge speed
HCC	quickly	slowly
HCS	Slowly	quickly

Table 2 shows that HCC and HCS both have their advantages and disadvantages. HCC responses quickly for the initial queries, but it converges slowly. HCS sorts result query each times, so its response time is very long for the initial queries. As the data merged to final partition is sorted, it converges much more rapidly.

To combine the advantage of both HCC and HCS, we apply a cost model. Based on the model, we decide which operation (cracking or sorting) to take in the query processing that could result in the best utility. We use a utility function: Utility = F(T), where T represents the response time. In the algorithm, we have three different ways to operate the data-- (a) sorting, (b) cracking and (c) binary search. Their computational complexity are $O(nlg_2n)$, $O(n)$ and $O(lg_2n)$ respectively.

When using cracking instead of sorting, we obtain a performance gain of F(TS)-F(TC) for the initial queries. In contrast, if we use sorting, we can obtain a performance gain of F(TC)-F(TB) for the subsequent queries. If F(TC)-F(TB)> F(TS)-F(TC), we will be better off by sorting the partition. We assume that the utility function F(T) is linear. Then the in equation can be rewritten to:

$$A(nlg_2n - n) > An - Blg_2n \qquad (1)$$

where A, B are constants, which mainly depends on the detailed implementation of the algorithms and the hardware environment.

Based on the above cost model, our algorithm for the hybrid adaptive index works as follows:

Input: the query range a and b
Output: the query results

Search (a , b)
1、   in the final partitions, find the data partition that contain a and b
2、   If the data partition is sorted, using binary search.
3、   else using cracking to find the data.
4、   If the cracked data could sort, sort it.
5、   If the result form the final partitions not answer the query
6、   find the query part in the initial partitions, using cracking.
        Merge the result into temp partition.
7、   If data in the temp partition can sort, sort it.
8、   Add the temp partition to final partitions.
9、   Return the results

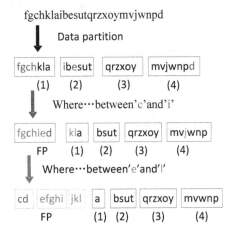

Fig. 5. Hyper Crack Sort\Crack

In Fig5, when the first query comes, it asks for the range between 'c' and 'i'. There is no final partition, so query in the initial partitions. Based on the cost model, the merged result "fgchied" could not sort, so just add it to the final partition. Then a new query asking for range "e – l" comes. It query in the final partitions. The partition that contains 'e' is unsorted, so cracking it to find the result. After cracking, the data partition is split into two parts "fghie" and "cd". Based on the cost model, the two partitions can be sorted, so the two partitions in the final partition are ordered. The result from the final partitions can't full answer the query. So it searches in the initial partitions. Based on the cost model, the merged result "klj" can be sorted, so sort the partition and add it to the final partitions.

From the example, it shows that hyper crack sort/crack algorithm response quickly in the initial query. For the first query, it only merges the result into final partitions, without sort. At the same time, In HCSC algorithm, data converges rapidly in the final partitions, only after two queries, the data in final partition is ordered.

4   Experimental Analysis

In this section, we demonstrate that HCSC algorithm is advantageous to both HCC and HCS.

Experiment Set-Up. We implemented our algorithms in C++.We used the C++ Standard Template Library for the cracker indices. All experiments ran on a 32-core hyper-threaded machine (4 Intel E5-2670 @2.6GHz)with 256GB RAM running SUSEOS 11 (64-bit). As in the previous work on adaptive indexing, our experiments were all main-memory resident, targeting on modern main-memory column-store systems.

We purposely kept the queries simple so we could isolate adaptive behavior using metrics such as (a) the response time for each individual query, (b) how fast performance reaches the optimal level, i.e., retrieval performance similar to a fully optimized index. The queries for experiments were of the following form.

select A from R where A > low and A < high

Analysis. In our experiment we used a column of 3×10^8 tuples with randomly distributed integer values in $[0,10^8]$.We fired 10^3 queries, where each query asked for a randomly located range with 10%,1%and 0.1% selectivity, respectively. This widely spread access pattern necessarily requires many queries to build-up index information. The data in the storage will eventually be sorted. We consider the two aspects: the query response time in the initial part and how many queries should be processed before the data converge to a complex index.

Lower Investment. Table 3 shows the three algorithms' response time for the first 10 queries. HCS responses slowly in the beginning, as it has to sort every result set from the initial partitions. It is clear that sorting is the major cost in the query processing. HCC response much more quickly, as it only needs to merge the result

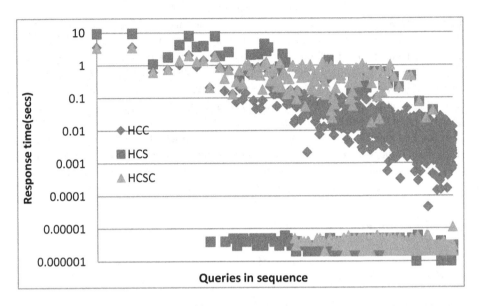

Fig. 6. The result of 10% selectivity

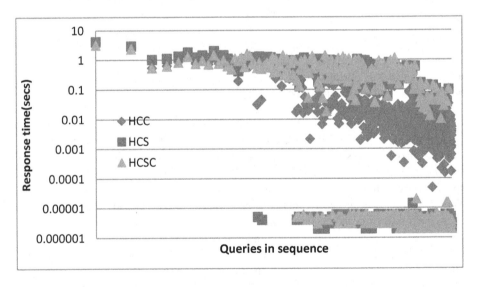

Fig. 7. The result of 1% selectivity

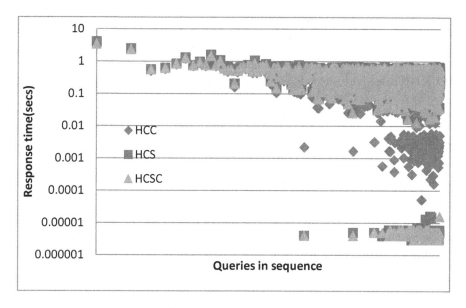

Fig. 8. The result of 0.1% selectivity

Table 3. The response time of three algorithm sunder 10% selectivity

	HCC	HCS	HCSC
1	3.406993	9.382217	3.57472
2	3.469169	9.488477	3.603923
3	0.60294	1.100447	0.603823
4	0.744287	1.793415	0.747324
5	1.039929	4.209944	1.442165
6	1.959421	7.904691	2.042448
7	0.904435	3.711497	1.302961
8	1.38513	3.907223	1.40169
9	0.205918	0.000004	0.206194
10	3.406993	9.382217	3.57472

sets from the initial partitions to the final partitions without performing any additional operations. HCSC uses a cost model to decide whether to perform sorting. Its response time is quicker than HCS and slightly slower than HCC

For the ninth query, HCS is several orders of magnitude faster than the other two algorithms, because the data queried is already in the final partitions and sorted. HCS uses binary search to identify the location of results, which is fast. In contrast, HCC must scan the final partitions. For HCSC, it depends on whether the partition was sorted previously.

Faster Adaption. Fig 6 shows the response time under the 10% selectivity. HCS converges rapidly. It only needs 43 queries to reach the completely ordered state. HCC maintains the smooth behavior of cracking, but it does not achieve fast convergence. HCSC uses a cost model to decide whether to sort the data. In the experiments, it requires 144 queries to construct the optimal index.

Effect of Selectivity. From Fig 6, Fig 7 and Fig 8, it shows that when the selectivity decreases, it takes more queries to reach the optimal performance. In addition, we observe that with smaller selectivity, the difference in convergence is less significant between the three algorithms. In Fig 8, under the 0.1% selectivity, HCSC shows similar performance to HCS.

5   Conclusion

In this paper, we propose a cost model for database cracking, and based on the model we designed a new algorithm HCSC.HCSC algorithm has the advantage of both HCC algorithm and HCS algorithm. With high selectivity, its advantage is more obvious. When the selectivity is small, HCSC's performance is close to that of HCS. Experiments show that the new algorithm responses quickly in the initial queries and converges rapidly.

Acknowledgement. This work is partly supported by the Important National Science & Technology Specific Projects of China ("HGJ" Projects, Grant No.2010ZX01042-001-002, 2010ZX 01042-002-002-03), the National Natural Science Foundation of China (Grant No. 61070054, 60873017,61170013), the Postgraduate Science & Research Funds of Renmin University of China under Grant No.12XNH177,and Basic Research funds of RUC, No. 12XNLJ01

References

1. Chaudhuri, S., Weikum, G.: Rethinking database system architecture: Towards a self-tuning risc-style database system. In: Proceedings of the 26th Int. VLDB, pp. 1–10 (2000)
2. Graefe, G., Idreos, S., Kuno, H., Manegold, S.: Benchmarking adaptive indexing. In: Nambiar, R., Poess, M. (eds.) TPCTC 2010. LNCS, vol. 6417, pp. 169–184. Springer, Heidelberg (2011)
3. Graefe, G., Kuno, H.: Adaptive indexing for relational keys. In: ICDE, pp. 69–74 (2010)
4. Graefe, G., Kuno, H.: Self-selecting, self-tuning, incrementally optimized indexex. In: EDBT, pp. 371–381 (2010)
5. Idreos, S., Kersten, M.L., Manegold, S.: Database cracking. In: CIDR, pp. 68–78 (2007)
6. Idreos, S., Kersten, M.L., Manegold, S.: Updating a cracked database. In: SIGMOD, pp. 413–424 (2007)
7. Idreos, S., Kersten, M.L., Manegold, S.: Self-organizing tuple reconstruction in column stores. In: SIGMOD, pp. 297–308 (2009)
8. Idreos, S., Manegold, S., Kuno, H., Graefe, G.: Merging what's cracked, cracking what's merged: adaptive index in main-memory column-stores. PVLDB 4(9), 585–597 (2011)

9. Halim, F., Idreos, S., Karras, P., Yap, R.H.C.: Stochastic database cracking:Towards robust adaptive indexing in main-memory Column-stores. PVLDB 5(6), 502–513 (2012)
10. Kersten, M., Manegold, S.: Cracking the database store. In: CIDR, pp. 213–224 (2005)
11. Bruno, N., Chaudhuri, S.: Physical design refinement: the'merge-reduce' approach. ACM TODS 32(4), 28:1–28:41 (2007)
12. Chaudhuri, S., Narasayya, V.R.: Self-tuning database systems: Adecade of progress. In: VLDB, pp. 3–14 (2007)
13. Finkelstein, S.J., Schkolnick, M., Tiberio, P.: Physical database design for relational databases. ACM TODS 13(1), 91–128 (1988)
14. Härder, T.: Selecting an optimal set of secondary indices. In: Samelson, K. (ed.) ECI 1976. LNCS, vol. 44, pp. 146–160. Springer, Heidelberg (1976)
15. Seshadri, A.: Generalized partial indexes. In: ICDE, pp. 420–427 (1995)
16. Stonebraker, M.: The case for partial indexes. SIGMOD Record 18(4), 4–11 (1989)
17. Hoare, C.: Algorithm 64: Quicksort. Comm. ACM 4(7), 321.0 4534 (1961)

Influence Maximization Algorithm Using Markov Clustering

Chungrim Kim, Sangkeun Lee, Sungchan Park, and Sang-goo Lee

School of Computer Science & Engineering
Seoul National University, Seoul 151-742, Korea
{merripu,liza183,baksalchan,sglee}@europa.snu.ac.kr

Abstract. Social Network Services are known as a effective marketing platform in that the customers trust the advertisement provided by their friends and neighbors. Viral Marketing is a marketing technique that uses the pre-constructed social networks to perform maketing with small cost while maximizing the spread. Therefore, which seed user to select is the primary concern in viral marketing. Influence maximization problem is a well known problem to find the top-k seed users who can maximize the spread of information in a social network.

Since obtaining the global optimal solution for the influence maximization problem is proven to be NP-Hard, many greedy as well as heuristic approach has been researched. However, greedy approaches take to much time to obtain the seed node, whereas the heuristic approaches show poor performance. To remedy such problems, we exploit the community structures in the social network to enhance the performance of the heuristic approaches. We perform markov clustering to find the natural communities in the social network and consider the most influential user in the community as the candidate for the top-k seeds. Also, we propose a novel attractor identification algorithm that finds the influential nodes in the community with reduced runtime, and 3 new hybrid approaches for influence maximization problem. Experiments show that the proposed algorithms are more scalable than the greedy approaches, whereas the influence spread obtained by those outperforms the heuristic approaches.

Keywords: Influence Maximization, Markov Clustering.

1 Introduction

Recently, Social Network Services(SNS) such as Facebook[1] or Twitter[2]are arising and its users are constantly increasing. Users of SNS services can interact with other users and form a community disregardin any temporal or geological constraints. SNS services serve as a medium to spread information, influence and ideas throughout the users, and therefore acknowled as a successful advertisement platform. With such tendency, viral marketing using pre-constructed social networks became a prominent figure.

---

[1] http://www.facebook.com
[2] http://www.twitter.com

B. Hong et al. (Eds.): DASFAA Workshops 2013, LNCS 7827, pp. 112–126, 2013.
© Springer-Verlag Berlin Heidelberg 2013

Viral marketing is a marketing methodology that uses the word-of mouth effect among the SNS users to perform advertisement about a specific product. For example, Hotmail[3] included an advertisment phrase saying "Get your private, free email at http://www.hotmail.com" in the e-mail of Hotmail users. With such advertisement, Hotmail could naturally spread the advertisement among the pre-constructed e-mail network. As the result of the viral marketing, Hotmail has gathered 1,200 users in 2 years[16]. Nowadays, social commerce companies such as groupon[4] use viral marketing to attract customers. Viral marketing aims to achieve maximum advertisement effect within a given budget. For efficient marketing outcome, it is needed to carefully select the seed users who will initiate the marketing process.

To resolve the problem, Influence Maximization Problem aims to find the k people who will maximize the marketing outcome when selected as seed users. Many researches until recently have proposed numerous algorithms for solving the Influence Maximization Problem, either by improving the greedy algorithm or proposing new heuristics. However, there are shortcommings in both approaches that the greedy algorithms' runtime are too large and the heuristics do not take the prominent community structures of the social network. Our contributions in this paper are in threefolds. First, we propose a novel heuristic approach that takes the community structures into account. Second, to further improve the runtime, we propose a novel algorithm to detect the influential node of each community without executing the whole graph clustering algorithm. Last, we propose two hybrid algorithms that combines the attractor detection algorithm with the existing greedy algorithm and the heuristics.

2   Problem Definition

2.1   Social Network Graph

The social network is represented as a weighted directed graph where nodes represent members of the social network and the edges represent relationships or interactions among them.

A weighted directed graph $G = (V, E)$ is comprised of tuples between the set of nodes, V, and the set of edges, E. An edge $e \in E$ can be represented as a pair of two nodes $u, v \in V$ $e = (u, v)$, and the direction from u to v. $e = (u, v)$ has $c_{u, v}$, the number of interaction between two nodes as weight. The set of neighbors of $u \in V$, $N_G(u)$, is defined as follows.

$$N_G(u) = \{ v\ inV \mid \exists (u,\ v) \in E \} \tag{1}$$

The weight of each edge is normalized by dividing each edge weights by the sum of weights.

$$Pr(v \| u) = \frac{(u,\ v).weight}{\sum_{w \in N_G(u)} (u,\ v).weight} \tag{2}$$

---

[3] http://www.hotmail.com
[4] http://www.groupon.com/

2.2   Information Spread Model

There are numerous information diffusion models to simulate the spread of information among a network. [1] [4] [10] [9] [13] discuss the "word-of-mouth effect" in the real world. Two of the most basic and widely-studied models will be considered in this paper. Firstly, [10] [9] [15] proposes the Independent Cascade(IC) Model. Each edge in the social network has same probability to influence the target node. The information diffusion under the IC model is simulated as follows.

Definition 1. *Every node can be either active or inactive. An active node represents an influenced user in the social network. The seed set of active nodes is defined as A_0. Newly activated nodes in the ith iteration are defined as A_i. in the $i + 1$th iteration, a node u in A_i tries to activate its inactive neighbor v with probabiltiy of $p_{u,\,v}$. when u successfully influences v, v is added to A_{i+1} and becomes active. Such iteration is repeated until $A_{i+1} = \emptyset$. The probability of u influencing v, $p_{u,\,v}$, is defined as follows.*

$$p_{u,\,v} = 1 - (1 - p)^{c_{u,\,v}} \tag{3}$$

p in the above formula represents the propagation probability, which is the probabilty of u influencing v with one interaction.

In the IC Model, nodes with high degree have high probability both to influence its neighbor and to be influenced by them. But in some application, nodes with high degree can be less influenced by its neighbor. For example, a person with 100 friends is not easily influenced by one of his friends. However, a person with only one friend can be easily influenced by his only friend. With such intuition, [11] proposed the Weighted Cascade(WC) Model.

Definition 2. *In the WC model, the probability of u influencing v, $p_{u,\,v}$, is defined as follows.*

$$p_{u,\,v} = \frac{c_{u,\,v}}{\sum_{i \in N_G(v)} c_{i,\,v}} \tag{4}$$

2.3   Influence Maximization Problem

Domingos and Richardson[7][18] were the first to define the Influence Maximization Problem as a probablistic algorithm problem. Kempe[11] defined the Influence Maximization Problem as an optimization problem, and proved that such problem is NP-Hard under the IC Model and the WC Model. The Influence Maximization Problem defined by Kempe[11] is as follows.

Definition 3. *Given a graph $G = (V, E)$ and the weights for each $(u,\ v) \in E$ representing the probability of u influencing v, the Influence Maximization Problem finds a set of nodes $S \subseteq V$ that maximizes the influence function $f(S)$ and $\|S\| = k$.*

3   Related Work

3.1   Greedy Approach

The greedy approach proposed by Kempe calculates the expected influence of each nodes using the Monte-Carlo simulation and adds the node with the highest expected value to the seed set, S. The result of the greedy algorithm guarantees the influence spread within (1 - 1 / e) of the optimal solution under IC and WC models[11]. Algorithm 1 describes the greedy alogithm.

Algorithm 1. GeneralGreedy(G, k)

Require: graph $G = (V, E)$, k for the number of seeds to be selected
Ensure: the set S that maximizes the influence spread
 1: $S = \emptyset$
 2: **for** i = 1 to k **do**
 3:    **for each** vertex $v \in V \backslash S$ **do**
 4:       $s_v = 0$
 5:       **for** i = 1 to R **do**
 6:          $s_v \mathrel{+}= |f(S \cup \{v\})|$
 7:       **end for**
 8:       $s_v = s_v / R$
 9:    **end for**
10:    $S = S \cup \{\arg\max_{v \in V \backslash S}\{s_v\}\}$
11: **end for**
12: **return** S

However, the greedy approach's runtime is slow because the expected influence for each nodes are calculated using Monte-Carlo simulation. Since the process of influence spread is defined with probabilistic models, the influence of a node can only be measured by simulating the influence spread multiple times and obtaining the approximate value. More runs of Monte-Carlo simulation can improves the approximation, but also takes more time to complete the Monte-Carlo simulations.

To remedy such shortcoming, [14] proposed a method to reduce the runtime by minizing the calculation of influence spread $f(u)$ of a node u. [14] uses a priority queue to recalculate the influnce spread of influential nodes (Cost-Effective Lazy Forward). Only the influence of the node with the highest influence is recalculated. When the recalculated influence spread is still greater than other nodes' influence spread, that node is added to the seed set. However, in the first iteration, influence spread of all nodes still need to be calculated. Therefore it is still inefficient in a large social network graph. [5] also proposed an algorithm that pre-eliminates the edges to calulate the influence spread of all nodes proportional to the node size for one Monte-Carlo Simulation. However, multiple Monte-Carlo Simulation need to be run to obtain more approximate influence spreads. The runtime increases as the number of Monte-Carlo Simulation increases.

3.2 Heuristic Approach

As the greedy algorithms' running time is still large despite the improvements introduced in previous sections, they may not be suitable for large social network graphs. Heuristic Approaches prove to be efficient alternatives for the Influence Maximization Problem.

The most basic approach is the degree centrality heuristic[19]. Intuitively, a user with many friends are influential in a social network. Using such intuition, the degree centrality heuristic selects k nodes that have the highest degree. This heuristic is frequently used in sociology to mine the most influential individual in a social network. [12] experimentally shows that the degree centrality heuristic outperforms other heuristics in the Influence Maximization problem. But the nodes selected by the degree centrality heuristic only consider its neighbors and therefore cannot be guaranteed to select the optimal seed set. When nodes with high degree are positioned nearly, the influence spread only affects nodes within a certain region.

To remedy such shortcomming, [5] proposes the degree discount heuristic. When a node u is selected as a member of the seed set, the degree of the nodes in $N_G(u)$ are discounted by one. In the next iteration, the node with the highest degree after the discount is selected as a member of the seed set. k iterations are performed to select k nodes with the highest degree. Although degree discount heuristic finds node with larger influence spread than the conventional degree centrality heuristic, it still disregards the community structure of the social network and therefore is apt to select nearby placed nodes. Lastly, [6] uses eigenvector centrality heuristic to select the k influential nodes. When a social network is represented as a transitional matrix, PageRank values for each node are calculated, and the k nodes with the highest PageRank values are selected as seed set.

Heuristic approaches are faster than the greedy approaches, but the seed set tend to be less influential, meaning that the influence spread of the obtained seed set is lower than those of the greedy approaches. Therefore we aim to improve the performance of the heuristic approaches.

4 Influence Maximization Using Markov Clustering

4.1 Markov Clustering

[8] first proposed the Markov Clustering which is a frequently used graph clustering algorithm in Bioinformatics. Markov Clustering divides the graph with a simple intuition. Assume that there exist multiple communities in a social network. When a k-step random walk is performed from a node in the graph, it is usual for the random walker to stop at one of the nodes within the community where u belongs to rather than nodes outside the community. Markov Clustering(MCL) uses such intuition and clusters the nodes whose random walker stops in the same node.

MCL performs two iterative operations repeatedly to find clusters of a graph. Each operations are named as Expansion and Inflation. One successive expansion and inflation operation is called as one iteration. MCL calculates the probability of a random walker stopping at a certain node using the expansion operation. The expansion operation multiplies the transition matrix of a social network graph with itself to calculate the transition probability with twice the random walk step as before. After the expansion operation, MCL uses inflation opration to speed up the conversion. Inflation operation increases the transitional probability of an edge with high weight, whereas decreases the transitional probability of an edge with low weight. Inflation operation modifies the transitional probability by firstly computing the transitional probability to the power of inflation rate. If the newly calculated value is below a certain threshold, that edge is removed and the whole transitional matrix is re-normalized. Expansion and Inflation operations are repeated until the transitional matrix becomes doubly idempotent, in other words, until the transitional matrix of ith iteration and i+1th iteration becomes identical.

The resulting transitional matrix contains the information about attractors and the nodes that are attached to the attractors. The column of the resulting transitional matrix are the starting nodes, whereas the rows are the result nodes. If a node u is attached to the attractor v, the value in the transitional matrix M[u][v] has a value larger than 0. Therefore, a row that contains more than one non-zero values is an attractor, and the number of nodes that are attached to the attractor is the size of the cluster.

4.2   Attractor Detection Using MCL

MCL's main aim is to divide a graph in multiple clusters. However in the Influence Maximization Problem, it is more important to obtain the attractors fastly, as the attractors are the most influential node in the cluster. Therefore we propose a novel algorithm that uses MCL to obtain the attractors fastly. MCL uses matrix multiplication for the expansion operation and therefore has a time complexity of $O(n^3)$ for each iteration. Furthermore, MCL has to perform multiple iterations repeatedly until convergence. When investigating the MCL process, the attractors are moslty identified in the early iterations, but has to complete the remaining iterations until convergence. As explained before, obtaining the attractors is more important in the Influence Maximization Prolem, the remaining iterations can be skipped when most of the attractors are already identified. Algorithm 2 shows the pseudo code for the attractor detection algorithm.

The attractor detection algorithm runs similar to the conventional MCL algorithm. It repeats Expansion and Inflation operations to find the attractors of each clusters. The MCL algorithm can be divided into two phases, namely the growing phase and the shrinking phase. The growing phase of the MCL algorithm is when the number of non-zero values of the transitional matrix increases, and the shrinking phase is when the number of non-zero values decreases. As the expansion operation performs random walks, the transitional probability can become from zero to non-zero when a node becomes reachable after additional

---

Algorithm 2. `AttractorDetection`(G, r)

Require: normalized graph $G = (V, E)$, inflation parameter r
1: $M = M(G)$
2: $GrowingPhase = true$
3: **repeat**
4:      $prevNNZ = nnz(M)$
5:      $M = M^2$
6:      **for** $i \in V$ **do**
7:          **for** $j \in V$ **do**
8:              $M[i][j] = M[i][j]^r$
9:          **end for**
10:         **for** **do**
11:             **if** $M[i][j] < \theta$ **then**
12:                 $M[i][j] = 0$
13:             **end if**
14:         **end for**
15:         **for** $j \in V$ **do**
16:             $M[i][j] = \frac{M[i][j]}{\sum_{k \in V} M[i][k]}$
17:         **end for**
18:     **end for**
19:     **if** nnz(M) < prevNNZ **then**
20:         $GrowingPhase = false$
21:         $AC \leftarrow diag(M)$
22:     **end if**
23: **until** $GrowingPhase == true$
24: $AC = \emptyset$
25: **for** $i = 0 to M.length$ **do**
26:     **if** M[i][i] > AC[i] **then**
27:         $AC \cup i$
28:     **end if**
29: **end for**
30: **return** AC

---

random walk steps. On the contrary, the Inflation operation eliminates non-zero values if it does not exceed the given threshold. When the number of the newly created edges exceeds the number of the eliminated edges, MCL algorithm is in the growing phase, and otherwise in the shrinking phase.

In the shrinking phase, the transitional probability towards the atrractors increases due to the fact that the Inflation operation increases the transitional probability of edges that already have high transitional probability. Also, the transitional probability of edges to the non-attractors decreases upon the repetition of the two operations. Therefore, if an self-looping edge's transitional probability increases in the shrinking phase, that node is likely to become an attractor. With such intuition, we proposed a novel algorithm that stores the transitional probability of self-loops in the last iteration of the growing phase, and compares it with the values of the first iteration of the shrinking phase. The nodes whose transitional probability increased are selected as the "Attractor

Candidate". The detected attractor candidates can be used as candidates of the influential nodes in the Influence Maximization Problem.

4.3 Hybrid Algorithms for Influence Maximization Problem

We propose a novel heuristics for the Influence Maximization Problem using the attractor detection algorithm. First, MCL heuristic selects k attractors with the biggest cluster sizes. Assuming that an attractor will influence most of the nodes in the cluster, selecting k attractors with the biggest cluster size can maximize the influence spread. Algorithm 3 shows the pseudo code for the MCL heuristic. The AttractorDetection function refers to the before-mentioned attractor detection algorithm.

Algorithm 3. `MCL Heuristic`(G, k, r)

Require: graph $G = (V, E)$, k for the number of seeds to be selected, inflation rate r
Ensure: the set S that maximizes the influence spread
 1: $S = \emptyset$
 2: $AC = Attractor Detection(G, r)$
 3: **for** $i = 1 to k$ **do**
 4:    select $u = \arg \max_v \{clusterSize(v) | v \in AC \backslash S\}$
 5:    $S = S \cup u$
 6: **end for**
 7: **return** S

Secondly, MCL Greedy heuristic applies the greedy algorithm only to the attractor candidates obtained by the attractor detection algorithm. Conventional greedy algorithm need to calculate the influence spread of each nodes and therefore have large runtimes. However, the MCL Greedy heuristic only calculates the influence spread of attractor candidates, and therefore can reduce the runtime. But as the number of attractor candidates increases, the runtime of MCL Greedy heuristic also increases due to the fact that it uses Monte-Carlo simulations to simulate the influence spread for the attractor candidates.

Lastly, the MCL Degree Discount heuristic combines the attractor detection algorithm and the degree discount heuristics which shows the best performance among the heuristics. MCL Degree Discount heuristic considers the community structure of the social network, but does not simulate the influence spread. Therefore it can achieve better performance while running faster than the conventional greedy algorithm.

5 Experiment

5.1 Datasets

Three datasets are used for the experiment. The 'High Energy Physics - Theory Collaboration Network' dataset proposed in [11] [12] [5] are the mostly used dataset in the literature. Also, 'Computational Geometry Collaboration Network'[2] is also used. Both datasets are co-authorship networks. Since real-world dataset of facebook or twitter are large in size and therfore greedy algorithm cannot be run on such datasets. However the co-authorship networks of various sizes are open to public and are known to imply the features of general social networks[17]. Both graphs have authors as nodes and edges when two authors have co-authored a paper. The weight of the edge is the number of papers that the two authors co-written. Each dataset will be referred to HEPT and GEOM in the following sections. Lastly a small-sized real world social network dataset is used to demonstrate the effect of the novel heuristices. This dataset consists of 9 communities and were open to public at NodeXL Graph Gallery[5]. This dataset will be referred as FB. The statistics of each dataset are as follows.

Table 1. Datasets

graph data name	# of nodes	# of edges
HEPT	15,233	58,891
GEOM	7,343	11,898
FB	367	3,728

5.2   Experiment Setup

We compare the proposed heuristics with the new greedy algorithm proposed in [5], degree discount heuristic, random selection, and lastly eigenvector centrality heuristic[3]. This experiment measures the influence spread of each algorithms for the datasets under the IC model and the WC model. The size of the seed set, k, is varied from 1 to 10. The restart probability for the eigenvector centrality heuristic is set to 15%. To measure the influence spread of each node, 1000 Monte-Carlo simulations are executed.

5.3   Effect of Attrator Detection Algorithm

In this experiment, we aim to show the effect of the attractor detection algorithm. Let us define the attractors obtained by fully executing the MCL algorithm as MCL, and the attractors obtained with the attractor detection algorithm as $eMCL$(early-terminated MCL). The recall and precision for the two datasets are calculated using Eq.5.

$$precision = \frac{eMCL \cap MCL}{eMCL}, recall = \frac{eMCL \cap MCL}{MCL} \qquad (5)$$

The precision and recall values for the HEPT and the GEOM datasets are as follows.

---

[5] http://www.nodexlgraphgallery.org/Pages/Graph.aspx?graphID=584

Table 2. precision & recall

graph data name	precision	recall
HEPT	0.7692	0.7135
GEOM	0.8495	0.8234

As the result of the experiment shows, attractors with about 80% precision and 76% recall in average are obtained with the attractor detection algorithm. This is due to the fact that the attractor detection algorithm only detects attractor candidates and not the precise clusters. The runtime of each algorithm are shown the table 3.

Table 3. runtime comparison (sec)

graph data name	MCL	eMCL
HEPT	1058.97	235.11
GEOM	118.73	29.76

eMCL in comparison to MCL terminates about 7.45 times faster in HEPT dataset and 4.87 times faster in GEOM dataset. When finding the top 10 nodes that maximizes the influence spread, 8 of the nodes obtained with eMCL were identical to the node obtained with MCL in HEPT dataset. In GEOM dataset, all 10 nodes were identical. The performance comparison between eMCL and MCL will be explained in the next experiment. Shown that the eMCL finds most of the attractors that MCL finds, it is shown that the attractor detection algorithm is efficient that it terminates faster than MCL.

5.4 Influence Spread Comparison

In this experiment, the size of the seed set k is varied under IC model and WC model to demonstrate the effectiveness of each algorithm. Also, runtime of each algorithms are compared for a given value of k which being 10.

Figure 1 shows the influence spread of each algorithm for the three datasets. For all datasets, random selection heuristic show the lowest influence spread and the eigenvector centrality heuristic show second lowest. Degree discount heuristic outperforms other heuristics, but shows slightly lower influence spread than the MCL, eMCL, eMLC Greedy and eMCL Degreee Discount heuristics. The conventional greedy algorithm shows similar influence spread as the proposed heuristics.

For the HEP dataset, the eMCL heuristic, eMCL Greedy heuristic, eMCL Degree Discount heuristic have influence spread that are slightly larger compared to the Degree Discount heuristics. The newly proposed heuristics influence about 95% of the nodes that are influenced by the greedy algorithm. For the GEOM dataset, the hybrid heuristics show 9.3% 20.7% increase in the influence spread.

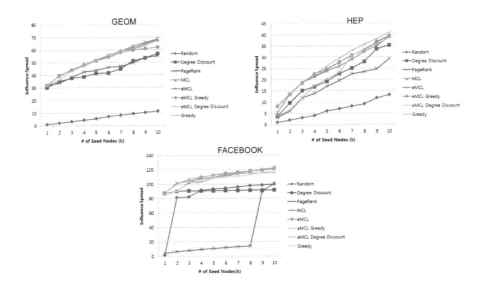

Fig. 1. Influence spread comparison under the IC Model

The eMCL Greedy heuristic influences about 92.1% compared to the greedy algorithm, whereas the eMCL Degree Discount and the eMCL heuristic slightly outperform the greedy algorithm. It can be also seen that eMCL heurisitic influences as much nodes as the MCL heuristic. The result for the facebook dataset proves that regarding the community structure in the Influence Maximization Problem can improve the influence spread in the social network. As the size of the seed set increases, the newly proposed heuristics select influential nodes that do not belong to the same community. When k is set to 10, eMCL Greedy and eMCL Degree Discount show similar influence spread as the greedy algorithm. MCL and eMCL heuristic shows 2.9% and 3% increase in the influence spread. However, the Degree Discount heuristic only shows 77.1% compared to the greedy algorithm.

Figure 2 show the influence spread of each algorithm for the three datasets. The overall result are similar to the experiment conducted on the IC model. For all datasets, random selection heuristic show the lowest influence spread and the eigenvector centrality heuristic follows. Degree discount heuristic show larger influence spread than other heuristics, but smaller compared to the MCL, eMCL, eMLC Greedy and eMCL Degreee Discount heuristics. The conventional greedy algorithm shows similar influence spread as the proposed heuristics. For the HEP dataset, it is shown that the newly proposed heuristics have influence spread that are about 7.9% 13% larger compared to the Degree Discount heuristics. The newly proposed heuristics influence about 94.8% of the nodes with the eMCL heuristic, and 96.9% with eMCL Greedy and eMCL Degree Discount compared to the greedy algorithm. It is shown that for the GEOM dataset the hybrid heuristics show 5%, increase in the influence spread in average. Under the

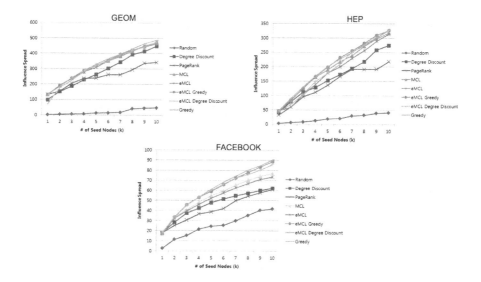

Fig. 2. Influence spread comparison under the WC Model

WC model, the newly proposed heuristics largely outperformed the degree discount heuristic in the GEOM dataset. All three heuristics showed about 96.9% of the influence spread compared to the greedy algorithm. Similar to the experiment under the IC model, regarding community strutures under the WC model improves the influence spread. For the Facebook dataset it is observable that regarding the community structure improves the influence spread. As the size of the seed set increases, the newly proposed heuristics' influence spread outperforms the conventional heuristics. eMCL Greedy shows similar influence spread as the greedy algorithm. eMCL Degree Discount heuristic shows about 94.9% influence spread, whereas the MCL and eMCL heuristics show 85.1% and 81.2% influence spread. The Degree Discount heuristic only shows 68.7% compared to the greedy algorithm.

5.5   Runtime Comparison

The greedy algorithm's runtime under the IC model takes longest to complete and the eMCL Greedy heuristics follows. Greedy algorithms runtime depends on multiple factors such as size of the graph, number of Monte-Carlo simulation, and the size of the seed set. As the size of the seed set, k, increases, the greedy algorithm will take longer to terminate. The eMCL and eMCL Degree Discount heuristics proposed in this paper terminates about 15 times faster than the greedy algorithm whereas their influence spread are similar to the greedy algorithm. The eMCL Greedy heuristic terminates 11 time faster than the greedy algorithm. The greedy algorithm's runtime also takes longest under the WC model. eMCL Greedy heuristics terminates 24.4 times faster than the greedy algorithm. Lastly, eMCL and eMCL Degree Discount heuristics terminates 46 times faster than the greedy algorithm in average. whereas their influence spread are similar to the greedy algorithm.

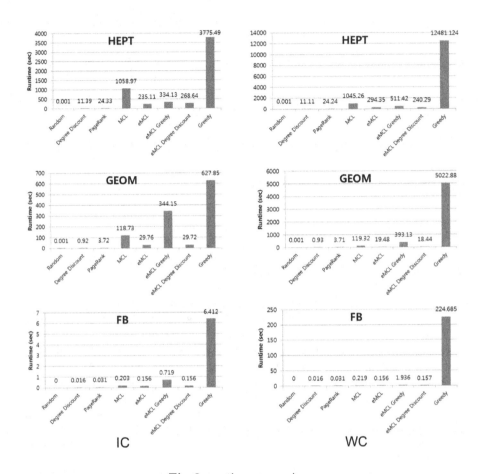

Fig. 3. runtime comparison

The last experiment is to show the scalability of each algorithm. Figure 4 shows the increase in runtime while varying the network size from 3000 to 15000. Simple heuristics such as Degree Discount heuristic or Random selection show almost insignificant increase in the runtime. However, greedy algorithm's runtime tend to drastically increase as the network size increases. On the contrary, the increase in the newly proposed heuristics are up to 4 times less than the greedy algorithm, and therefore is more scalable. As the network size increases, hybrid heuristics can handle larger networks than the greedy algorithm in limited timespan.

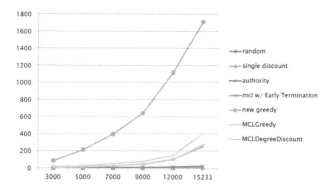

Fig. 4. comparison of runtime varying the network size

6   Conclusion

In this paper, we proposed novel heuristics for the Influence Maximization Problem that regards the inherent community structures in a social network. Also, we proposed an efficient algorithm that only selects the influential nodes in each communities as candidate nodes for the seed set. The efficiency of the attractor detection algorithm is experimentally shown in the experiment section. Using the attractor detection algorithm, we propose three hybrid heuristics for the Influence Maximization Problem. Our heuristics are advantageous in the means that it is more scalable than the conventional greedy algorithm, whereas shows larger influence spread than currently existing heuristics.

There are several future directions for this reserach. First, if MCL can be run in parallel, the scalability of the proposed heuristics will also improve. Extending the MCL to run parallely will be one of the directions. Secondly, extending the attrator detection algorithm to let the user choose its termination point. For example, a user might want more precise attractor candidates while sacrificing the runtime or vice versa. This extention would allow the user to choose what he/she values more, either the performance of the algorithm or the runtime.

Acknowledgments. This work was supported by the National Research Foundation of Korea(NRF) grant funded by the Korea government(MEST) (No. 20110017480).

References

1. Bass, F.M.: A new product growth for model consumer durables. Management Science 15(5), 215–227 (1969)
2. Batagelj, V., Mrvar, A.: Pajek datasets (2006)
3. Brin, S., Page, L.: The anatomy of a large-scale hypertextual web search engine. In: Proceedings of the Seventh International Conference on World Wide Web 7, WWW7, pp. 107–117. Elsevier Science Publishers B. V., Amsterdam (1998)

4. Brown, J.J., Reingen, P.H.: Social ties and word-of-mouth referral behavior. Journal of Consumer Research 14(3), 35–62 (1987)
5. Chen, W., Wang, Y., Yang, S.: Efficient influence maximization in social networks. In: Proceedings of the 15th ACM SIGKDD International Conference on Knowledge Discovery and Data Mining, KDD 2009, pp. 199–208. ACM, New York (2009)
6. Chen, W., Yuan, Y., Zhang, L.: Scalable influence maximization in social networks under the linear threshold model. In: ICDM, pp. 88–97 (2010)
7. Domingos, P., Richardson, M.: Mining the network value of customers. In: Proceedings of the Seventh ACM SIGKDD International Conference on Knowledge Discovery and Data Mining, KDD 2001, pp. 57–66. ACM, New York (2001)
8. Dongen, S.: A cluster algorithm for graphs. Tech. rep., CWI (Centre for Mathmatics and Computer Science), Amsterdam, The Netherlands, The Netherlands (2000)
9. Goldenberg, J.: Using complex systems analysis to advance marketing theory development: Modeling heterogeneity effects on new product growth through stochastic cellular automata. Academy of Marketing Science Review 9, 1–8 (2001)
10. Goldenberg, J., Libai, B., Muller, E.: Talk of the network: A complex systems look at the underlying process of word-of-mouth. Marketing Letters (2001)
11. Kempe, D., Kleinberg, J., Tardos, É.: Maximizing the spread of influence through a social network. In: Proceedings of the Ninth ACM SIGKDD International Conference on Knowledge Discovery and Data Mining, KDD 2003, pp. 137–146. ACM, New York (2003)
12. Kempe, D., Kleinberg, J., Tardos, É.: Influential nodes in a diffusion model for social networks. In: Caires, L., Italiano, G.F., Monteiro, L., Palamidessi, C., Yung, M. (eds.) ICALP 2005. LNCS, vol. 3580, pp. 1127–1138. Springer, Heidelberg (2005)
13. Kimura, M., Saito, K.: Tractable models for information diffusion in social networks. In: Fürnkranz, J., Scheffer, T., Spiliopoulou, M. (eds.) PKDD 2006. LNCS (LNAI), vol. 4213, pp. 259–271. Springer, Heidelberg (2006)
14. Leskovec, J., Krause, A., Guestrin, C., Faloutsos, C., VanBriesen, J., Glance, N.: Cost-effective outbreak detection in networks. In: Proceedings of the 13th ACM SIGKDD International Conference on Knowledge Discovery and Data Mining, KDD 2007, pp. 420–429. ACM, New York (2007)
15. Lopez-Pintado, D.: Diffusion in complex social networks. Games and Economic Behavior 62(2), 573–590 (2008)
16. Montgomery, A.L.: Applying quantitative marketing techniques to the internet (2001)
17. Newman, M.E.J.: The structure of scientific collaboration networks. Proceedings of the National Academy of Sciences of the United States of America 98(2), 404–409 (2001)
18. Richardson, M., Domingos, P.: Mining knowledge-sharing sites for viral marketing. In: Proceedings of the Eighth ACM SIGKDD International Conference on Knowledge Discovery and Data Mining, KDD 2002, pp. 61–70. ACM, New York (2002)
19. Wasserman, S., Faust, K.: Social Network Analysis: Methods and Applications (Structural Analysis in the Social Sciences), vol. 63. Cambridge University Press (1994)

An Approximate Approach for Batch Update of Common Preference Group

Dawen Jia[1], Zhiyong Peng[2], Cheng Zeng[1], and Deqiang Cao[2]

[1] State Key Laboratory of Software Engineering, Wuhan University, China
[2] Computer School, Wuhan University, China
{brilliant,peng,zengc,d.q.cao}@whu.edu.cn

Abstract. There are a large amount of active Web users whose behaviors reveal their preferences. The user preference model can be established via analyzing user behaviors. Mining common preference group (CPG) from the user preference model can support many applications such as potential community discovery, data sharing and recommendation. However, the user preference model is constantly evolving, which is caused by users' various behaviors. With massive users, each time the user preference model changed, mining CPG from the scratch is not an option. In this paper, we analyze how users' behaviors influence CPG. Then, an approximate approach for batch CPG update is proposed to avoid CPG re-computing from sketch when the user preference model changed.

Keywords: User preference model, common preference group, frequent closed itemsets, formal concept analysis, incremental mining.

1 Introduction

In the circumstance of massive user generated unstructured data, the data sharing and recommendation approaches take more important role than information retrieval approaches for data diffusion. In literature [1], we analyzed the disadvantages of current data sharing and recommendation methods and proposed an automatic group mining approach based on user's common preferences, which lead to sufficient data diffusion and improve the sociability between users.

In the social Web, users have their own preferences about data objects. Currently, the systems provide some functions which allow users express their preferences, such as marking "favorite" or "like" to an object or give an object a rating score. Several semantic topics could be extracted from each object. The users' behaviors could reveal their preferences on each semantic topic. We have a hypothesis that a user like a data object because the user is interested in some semantic topics implied in the object. With this assumption, we switch interests of users from the objects to the semantic topics, and we group users who share common interests together as a *common preference group* (CPG). Intuitively, the essential of the grouping idea is that users who have the same interestingness towards a set of semantic topics should be gathered together as a CPG from the *user preference model* (UPM).

B. Hong et al. (Eds.): DASFAA Workshops 2013, LNCS 7827, pp. 127–138, 2013.
© Springer-Verlag Berlin Heidelberg 2013

Generally, more than one semantic topic could be extracted from a data object. *TopicSpace* denotes all possible topics that could be predefined or extracted from all objects. o_{topic} is a set of semantic topics $\{t_i\}$ implied by the object o.

$$o_{topic} \subseteq TopicSpace \tag{1}$$

A user u may interact with a set of objects. *ObjectSpace* denotes all objects. $u_{interaction}$ is a set of objects $\{o_i\}$ which user u have interacted with.

$$u_{interaction} \subseteq ObjectSpace \tag{2}$$

Based on (1) and (2), the users' preference on semantic topics can be derived. $u_{preference}$ is used to describe users' preference on a set of semantic topics. v_i is the user's interestingness on each semantic topic t_i.

$$u_{preference} = \{< t_i,\ v_i >\} \tag{3}$$

Definition 1: User Preference Model (UPM)

A *user preference model* (UPM) includes four components $< \mathbb{U},\ \mathbb{T},\ \mathbb{V}, \mathbb{P}>$. $\mathbb{U}$ refers to a set of users $\{u_1, u_2, ..., u_m\}$, $\mathbb{T}$ is a set of semantic topics $\{t_1, t_2, ..., t_n\}$ which should be consistent with user's interests. $\mathbb{V}$ is a set $\{V_1, V_2, ..., V_n\}$ in which each element V_i has a range $\{v_1^i, v_2^i, ..., v_{k(i)}^i\}$ representing $k(i)$ different levels of users' interestingness on t_i. Each user u could have only one interestingness level v^i in each semantic topic t_i, then $<t_i, v^i>$ is called a preference of user u. Let $\mathcal{P}(u)$ represents all preferences of the user u, the *User Preference Model* is $\mathbb{M}$, then

$$\mathbb{M} = \bigcup_{u_i \in \mathbb{U}} < u_i, \mathcal{P}(u_i) > \tag{4}$$

$\mathbb{P}$ is all preferences of all the users in $\mathbb{M}$, then

$$\mathbb{P} = \bigcup_{u_i \in \mathbb{U}} \mathcal{P}(u_i) \tag{5}$$

Given an UPM $\mathbb{M}$, let $U_k = \{u_1, u_2, ..., u_k\}$ is a subset of $\mathbb{U}$, then the common preferences of U_k is $P_k = \mathcal{P}(u_1) \cap \mathcal{P}(u_2) \cap ... \cap \mathcal{P}(u_k)$. Therefore, we could calculate the common preferences of any subset of $\mathbb{U}$, but not all the subsets of $\mathbb{U}$ could form a CPG. The concept of *common preference group* is based on the two following functions[1]:

$$f_{preference}(U) = \{p \in \mathbb{P} \mid \forall u \in U, p \sqsubset u\} \tag{6}$$

$$g_{user}(P) = \{u \in \mathbb{U} \mid \forall p \in P, p \sqsubset u\} \tag{7}$$

The function $f_{preference}(U)$ returns a set of preferences included by all the users belonging to U, while function $g_{user}(P)$ returns the set of users all of whom have a set of common preference P.

---

[1] '$p \sqsubset u$' denotes that the user u has the preference p.

Definition 2: Common Preference Closure (CPC)

A set of preferences P is said to be a *common preference closure* if and only if

$$\mathcal{C}_{cpc}(P) = f_{preference}(g_{user}(P)) = P \qquad (8)$$

The function $\mathcal{C}_{cpc}(P)$ is called *closure operator*, which returns a CPC that includes the set of preferences P. Two set of preferences belong to the same CPC if and only if they are contained by the same users. A set of preferences P is a CPC if no superset of P with the same users exists.

Definition 3: Common Preference Group (CPG)

Given a UPM, many CPC could be computed according to previous definitions. Each CPC $\mathcal{C}$ corresponds to a *common preference group* (CPG) g, which satisfies the condition that all users in the CPG g sharing the same set of preferences $\mathcal{C}$. Given a CPG g containing a user set $\mathcal{U} = \{u_1, u_2, \ldots, u_s\}$, their common preferences $\mathcal{H} = \mathcal{P}(u_1) \cap \mathcal{P}(u_2) \cap \ldots \cap \mathcal{P}(u_s)$. $\mathcal{U}$ and $\mathcal{H}$ are called *group user relation* (GUR) of g and the *group preference relation* (GPR) of g, respectively. The users in $\mathcal{U}$ are called the members of the CPG g which have the common preferences $\mathcal{H}$. $\mathcal{H}$ must be a CPC.

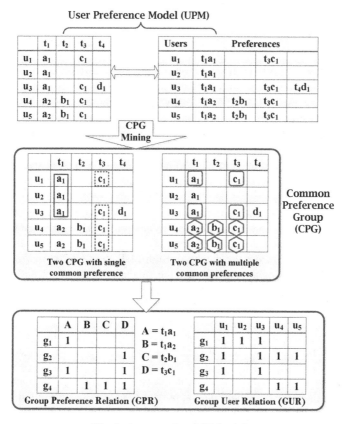

Fig. 1. An example of *CPG* mining

Definition 4: CPG Mining

The process of CPG Mining is discovering all CPG from a given UPM. Usually, we need to set the threshold *minUsers* denoting the minimal requirement of users to form a CPG. With a CPG g, its GPR and GUR are easily inferred. GUR indicates the relationship between users and CPG, namely group members. GPR indicate the relationship between user's preference and CPG, namely the common preference shared by all members in a CPG.

We give a simple example to demonstrate CPG mining. As depicted in Fig.1, we define the user set $\mathbb{U} = \{u_1, u_2, u_3, u_4, u_5\}$, the semantic topic set $\mathbb{T} = \{t_1, t_2, t_3, t_4\}$, and *minUsers* = 2. There are four CPG generated from this example. There are two CPG with single common preference and the other two CPG with multiple common preferences.

However, the user preference model is constantly evolving. New users continuously join the system and old users may de-register. Moreover, user's preferences are not static. In a word, the UPM is changing from time to time. With massive users, each time the user preference model changed, mining CPG from the scratch is not an option. An efficient CPG update approach is urgent if we want to keep them up to date. In this paper, an efficient batch updating approximate approach is proposed to avoid CPG re-computing from sketch when the user preference model changed.

2   Related Work

As far as we know, there is no research on mining CPG currently. The idea of CPG mining in this paper comes from the traditional frequent itemsets mining (FIM) and formal concept analysis (FCA)[10]. However, FIM ignores the same item confliction in the mining procedure and only considers the existing of items instead of their levels. In different purpose, FCA pays attention on those more representative itemsets and their inclusion relation between each other, so that the accuracy of the result is much lower than the requirement of CPG mining. For easy to understand and showing the differences of our approach, this section will introduce related works about the updating of FIM and FCA.

The research on FIM updating starts with literature [2] which proposed a Fast Update Algorithm (FUP) to update the frequent itemsets in a database when transactions are added to the database. Literature [3] improves FUP for the situation that an existing transaction is deleted from the database. Both FUP and FUP2 mine frequent itemsets by repeatedly scanning a database once the database has been updated.

Literature [4] proposed a new FIM updating algorithm based on negative border which is a collection of all infrequent itemsets. The method needs to scan the whole database once if the database update brings new infrequent itemsets to the negative border. However, this method is quite memory consuming. The Sliding-Window Filtering approach proposed in literature [5] first partitions a transaction database into several sequent partitions. Then it uses the minimum support in each partition to avoid generating unnecessary candidate itemsets. Consequently, the method has to scan the whole database to update the support of the frequent itemsets mined before instead of only scanning the changed partition of the database. So the two algorithms are low efficiency.

There are other algorithms try to minimize the calculated amount when the database changes. Literature [6] presented an algorithm called Update With Early Pruning that employed a dynamic look-ahead strategy. This strategy was used in updating the existing frequent itemsets by detecting and removing infrequent itemsets when new transactions were added to the original database. As a result, the number of candidate itemsets is minimised in an updated database. Literature [7] proposed the DB-tree and PotFP-tree data structures, which are the extensions of the classical FP-tree. Based on the two special data structures, the required number of database scans for incremental mining frequent itemsets is reduced.

Moreover, some researches focus on FIM of real-time data streams or using distributed technique for FIM updating. Literature [8] uses a parallel incremental approach at each site to produce and update a local model, then shares these distributed models at each site to produce and update the global model by minimising the communication cost. Literature [9] presented a novel incremental algorithm for mining closed itemsets over data streams. It uses the Closed Enumeration Tree to maintain a dynamically selected set of itemsets.

For updating based on FCA, literature [11] introduced an ontology construction method based on FCA and incremental calculation is considered in the paper. Literature [12] also presented an incremental method for concept formation in real world and FCA is the core of organizing concept relationship.

3   CPG Updating

New users continuously join the system and old users may de-register. Moreover, user's preferences are not static. In a word, the UPM is changing from time to time. With massive users, each time the user preference model changed, mining CPG from the scratch is not an option. An efficient CPG update approach is urgent if we want to keep them up to date.

Table 1. Relationships between user behavior and UPM modification

User Behavior	UPM Modification
A new user registers	Add a new line
An old user de-registers	Delete a line
A user's preferences change	Delete a line and then add a new line

The types of user behavior and their corresponding modification of UPM are given in Table 1. All user behaviors lead to two types of UPM modification: delete a line and add a new line. When an old user de-registers from the system, a line should be deleted from UPM; and when a new user registers to the system, a new line would be added to UPM. We can treat a user's preference change as deleting this user first and add a new user with the changed preferences at the same time. Therefore, the corresponding UPM modifications are deleting a line and then adding a new line to UPM.

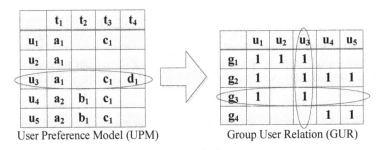

Fig. 2. User Deletion

Here, we discuss how these two types of UPM modifications influences CPG. When a user de-registers from the system, in other words, this user would be deleted from UPM, the following process is simple. At first, we check the group user relation (GUR), and delete the column of this user, and then we check if the CPG which delete the user still have enough users to make it valid. An example is given in Figure 2. The user u_3 is deleted from UPM, so we delete u_3 in UGR and if the minimal requirement of users is set to 2, then CPG g_3 should be removed from GUR. Obviously, batch deletion is more efficient since we only need to do the valid check once.

We already know the process of CPG update is quite simple when a line is deleted from UPM. However, the process would be much more complicated when a new line is added to the UPM. There are following situations.

***Situations* (1):** The preferences of the new coming user match with current CPG, then no new CPG will be generated, and we only need to add this user in corresponding CPG. An example is depicted in Fig.3.

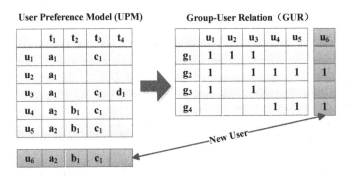

Fig. 3. Situation 1: user preferences matched

***Situations* (2):** The preferences of the new coming user partially match with current CPG. According to *definition 2*, a new CPG will be generated. An example is depicted in Fig.4.

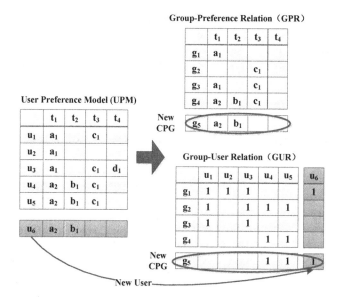

Fig. 4. Situation 2: user preferences partially matched

Situations (3): The new CPG are generated, because the new coming user makes the total number of the users reach the minimal requirement *minUsers*. As shown in Fig.5, only u_3 have preferences combination $\{t_3c_1, t_4d_1\}$, and this information is unclear. Since we have set the constraint of minimal users during CPG mining, those combinations whose users are less than *minUsers* are not recorded. It's not practical to record all of them.

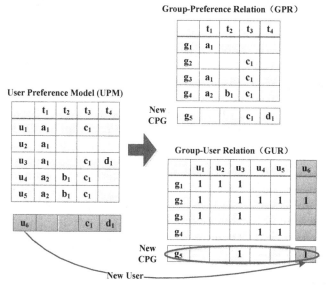

Fig. 5. Situation 3: user preferences mismatched

Updating CPG once a new line is added in UPM each time is not only unnecessary but also inefficient. Situation 1 and situation 2 are similar, and they rely on existed common preferences. In this paper, we propose an incremental approximate approach to realize CPG batch update, which can deal with situation 1 and situation 2. We will elaborate our approach in the following section and figure out how to deal with situation 3 in future.

4   An Approximate Approach for Batch CPG Update

As time goes on, we can treat the new coming lines as a new UPM *upmNew*. We can adopt the CPG mining algorithm again on *upmNew* to generate another set of CPG and we should set the minimal user requirement to one user. Since the size of *upmNew* is relatively small, the efficiency could be acceptable. Now, the problem is merging two set of CPG as the final result. From Definition 2 and Definition 3, we know that CPC is equivalent to GPR. If we know all the CPC, we can calculate the users whose preferences contain those CPC, and then we have all GUR. Therefore, there are two main steps to merge two different CPG:

(1) the first step is finding all the CPC;
(2) the second step is calculating the users who correspond to those CPC, also known as group user relation.

4.1   Calculate the CPC

In this subsection, we discuss how to find all the possible CPC when we merge two CPG.

Lemma 1. Given two different UPM, two set of CPC X_{CPC} and Y_{CPC} are mined from the two UPM, respectively. When we merge X_{CPC} and Y_{CPC}, the result is a set of CPC Z_{CPC}, and Z_{CPC} has following characteristics:

(1) $X_{CPC} \cup Y_{CPC} \subseteq Z_{CPC}$;
(2) Suppose N_{CPC} is the set of new generated CPC by the merging operation, $N_{CPC} = Z_{CPC} - (X_{CPC} \cup Y_{CPC})$, then $\forall$ CPC $\alpha \in N_{CPC}$, $\exists$ (CPC $\beta \in X_{CPC}$ & CPC $\gamma \in Y_{CPC}$), $(\alpha \subset \beta)$ and $(\alpha \subset \gamma)$.[2]

In our case, suppose that ORI_{CPC} and UPD_{CPC} are the CPC set mined from original UPM and the newly formed UPM, respectively. For the original UPM, the minimal user requirement to form a CPG is *minUsers*; for the newly formed UPM, the minimal user requirement is one user. According to Lemma 1, the following facts can be inferred: a) the ORI_{CPC} still exist in the final results; b) the UPD_{CPC} from the incremental part will not be always in final results, which depend on how many users having those common preference; 3) suppose that the new generated CPC set is New_{CPC}, for any CPC *newcpc* in New_{CPC}, in both ORI_{CPC} and UPD_{CPC}, we can find at least one CPC whose preferences is the superset of preferences of *newcpc*.

We design an efficient CPG merging algorithm based on *Maximum Common Preference closure (MCPC)*.

---

[2] CPC is a set of preference, therefore, $\alpha \subset \beta$ represents the preferences of α is the proper subset of the preferences of β.

Definition 5: Maximum Common Preference Closure (MCPC)

Let $\mathcal{C}_{cpc}$ denote all CPC in a given user UPM M, a CPC C is said to be a *maximum common preference closure* if and only if C is not the proper subset all other CPC. Therefore, MCPC(M) = { $C \mid C \in \mathcal{C}_{cpc}$, $\forall$ $X \in (\mathcal{C}_{cpc} - C)$, $C \not\subset X$ }.

Lemma 2. Given two different UPM M_1 and M_2, if $C_1 \in$ MCPC(M_1), $C_2 \in$ MCPC(M_2), $C = C_1 \cap C_2$, C is a preference set, then
(1) C must be a common preference closure;
(2) Any proper subset of C might be a common preference closure, if C' is the proper subset of C and C' is a common preference closure, then $|g_{user}(C')| > |g_{user}(C)|$.

According to above definition and lemma, we have the following conclusion.

Corollary 1.   For two sets of CPC X_{CPC} and Y_{CPC} are mined from the two different UPM X_{UPM} and Y_{UPM}, respectively. The result of merging X_{CPC} and Y_{CPC} is Z_{CPC}, then CPC in Z_{CPC} includes:
(1) all CPC in X_{CPC};
(2) a subset of CPC in Y_{CPC} depending on the number of users who have common preference Y_{CPC};
(3) any intersection k between elements in MCPC(X_{UPM}) and elements in MCPC(Y_{UPM});
(4) the non-empty proper subset of k is also the candidate CPC.

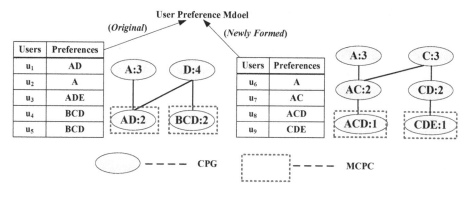

Fig. 6. CPG merging example

With corollary 1, we can get all possible CPC in the final result, and the next step is computing the users corresponding to each CPC, then we can get all valid CPC. We design a two level index structure to calculate the users of CPC, which will be elaborated in the next subsection.

Table 2. MCPC intersection

MCPC from original UPM:	AD, BCD
MCPC from newly formed UPM:	ACD, CDE
The results of MCPC intersection:	AD, D, CD

Here we use an example to demonstrate our CPG merging method. As depicted in Fig. 6, two UPM are given. CPG and MCPC are mined from the two UPM, respectively. The result of MCPC intersection is shown in Table 2. According to Corollary 1, the direct result of MCPC intersection is the valid CPC and their subset is the candidate CPC. However, given a set S whose number of elements is α, then the number of its non-empty proper subset is $2^{\alpha}-2$. According to the CPC definition, a set of preferences P is a CPC if no superset of P with the same users exists. Therefore, not all candidate CPC are valid. Since we only need to know the users who correspond to valid CPC, a pruning strategy is designed to reduce calculating the users who correspond to invalid CPC whenever possible. Here, Lemma 3 gives the pruning strategy.

Lemma 3. Given a valid CPC containing a set of preference P which correspond to a user set U, and a candidate CPC containing preference set P', which correspond to a user set U', is a proper subset of P, there are the following conclusion:

    a) if $U \subset U'$, then P' is a valid CPC;

    b) if $U = U'$, then P' and all its superset can be pruned.

For example, we suppose that the results of MCPC intersection include $ACDE$, then, $ACDE$ is a valid CPC. If the users who have preference A is the same as the users who have preference $ACDE$, then the all the candidate CPC containing preference A can be pruned (In this case, candidate CPC A, AC, AD, AE, ACD, ACE, ADE can all be pruned).

4.2   Calculate the Group User Relation

When we get all candidate CPC, the next step is calculate its users. The users of the candidate CPC are from both original and newly formed UPM. Since the users are different from the two UPM, we just need to combine the users together as the final GUR. We design a two level index to get the corresponding users of each CPC.

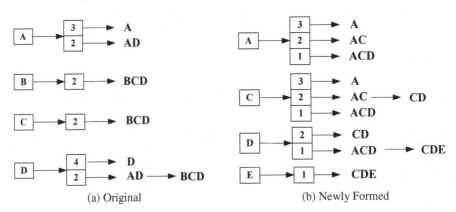

(a) Original   (b) Newly Formed

Fig. 7. Two level index

For all CPG mined from a UPM, we create a two level index for them. The first level is from preference to user's number, and the second level is from user's number to CPC. As shown in Fig. 7(a), preference A is corresponding to two CPC A and AD.

CPC A is corresponding to three users, and CPC AD is corresponding to two users. In this case we can calculate the users of any candidate CPC by this index. Take AD as an example. In the final CPG merging result, AD is a valid CPC. Its users come from both original and the newly formed UPM and we just need to combine the users together. The key problem is how to find the users in each UPM. Our method is intersecting the user number of A and the user number of D in two level index, and then sort those intersection numbers in descending order. We choose the largest number whose corresponding CPC contains AD as the result. In Fig. 7(a), intersection number of A and D is 2, as shown in Fig.6, both u_1 and u_3 have preferences A and D. In Fig. 7(b), the intersection numbers are 2 and 1, however, as shown in Fig.6, only one user u_8 has preferences A and D. Therefore, after CPG merging, CPC AD has corresponding users $\{u_1, u_3, u_8\}$.

5   Conclusions and Future Work

User's preferences are not static, so the UPM is changing from time to time. With massive users, each time the user preference model changed, mining CPG from the scratch is not an option. In this paper, we analyzed the types of user behavior and their corresponding modification of UPM, and then we summarized how the two types of UPM modifications influenced CPG. We proposed an approximate batch CPG update approach to keep them up to date and also avoid CPG re-computing.

When the new users are added into UPM, there are three situations. In this paper, we solve the problem caused by situation 1 and situation 2 together.   In the next step, we will figure out how to deal with situation 3. Then, we combine the result of these two steps and the final result should be identical with CPG re-computing approach. Meanwhile, we should make sure that time efficiency is also greatly improved.

Acknowledgements. This research is supported by the State Key Program of National Natural Science of China under Grant No. 61232002, the National Natural Science Foundation of China under Grant No. 61070011, 61202034, the Research Fund for the Doctoral Program of Higher Education No. 20110141120033, and the National High Technology Research and Development Program of China under Grant No. 2012AA011004.

References

1. Jia, D., Zeng, C., Nie, W., Li, Z., Peng, Z.: A New Approach for Date Sharing and Recommendation in Social Web. In: Liddle, S.W., Schewe, K.-D., Tjoa, A.M., Zhou, X. (eds.) DEXA 2012, Part II. LNCS, vol. 7447, pp. 314–328. Springer, Heidelberg (2012)
2. Cheung, D.W., Han, J., Ng, V.T., Wong, C.Y.: Maintenance of Discovered Association Rules in Large Databases: An Incremental Updating Technique. In: Proceedings of the Twelfth International Conference on Data Engineering (ICDE 1996), pp. 106–114. IEEE Computer Society (1996)

3. Cheung, D.W., Lee, S.D., Kao, B.: A general incremental technique for maintaining discovered association rules. In: Topor, R.W., Tanaka, K. (eds.) DASDAA 1997. LNCS, vol. 6, pp. 185–194. Springer, Heidelberg (1997)
4. Thomas, S., Bodagala, S., Alsabti, K., Ranka, S.: An efficient algorithm for the incremental updation of association rules in large databases. In: Proceedings of the Third International Conference on Knowledge Discovery and Data Mining (KDD 1997), pp. 263–266. AAAI Press (1997)
5. Lee, C., Lin, C., Chen, M.: Sliding-window filtering: an efficient algorithm for incremental mining. In: Proceedings of the Tenth International Conference on Information and Knowledge Management (CIKM 2001), pp. 263–270. ACM Press (2001)
6. Ayan, N.F., Tansel, A.U., Arkun, E.: An efficient algorithm to update large itemsets with early pruning. In. Proceedings of the Fifth International Conference on Knowledge Discovery and Data Mining (KDD 1999), pp. 287–291. AAAI Press (1999)
7. Ezeife, C.I., Su, Y.: Mining incremental association rules with generalized FP-tree. In: Cohen, R., Spencer, B. (eds.) Canadian AI 2002. LNCS (LNAI), vol. 2338, pp. 147–160. Springer, Heidelberg (2002)
8. Otey, M., Veloso, A., Wang, C., Parthasarathy, S., Meira, W.: Mining frequent itemsets in distributed and dynamic databases. In: Proceedings of the Third IEEE International Conference on Data Mining (ICDM 2003), pp. 617–620. IEEE Computer Society (2003)
9. Chi, Y., Wang, H., Yu, P.S., Muntz, R.R.: Moment: maintaining closed frequent itemsets over a stream sliding window. In: Proceedings of the Fourth IEEE International Conference on Data Mining (ICDM 2004), pp. 59–66. IEEE Computer Society (2004)
10. Bernhard, G., Rudolf, W.: Formal Concept Analysis: Mathematical Foundations. Springer-Verlag New York, Inc. (1999)
11. Peng, X., Zhao, W.: An Incremental and FCA-Based Ontology Construction Method for Semantics-Based Component Retrieval. In: Proceedings of the Seventh International Conference on Quality Software (QSIC 2007), pp. 309–315. IEEE Computer Society (2007)
12. Maddouri, M.: Towards a machine learning approach based on incremental concept formation. Journal of Intelligent Data Analysis 8(3), 267–280 (2004)

A Probabilistic Approach for Events Identification from Social Media RSS Feeds

Chiraz Trabelsi[1] and Sadok B. Yahia[1,2]

[1] Faculty of Sciences of Tunis, University Tunis El-Manar, 2092 Tunis, Tunisia
[2] Department of Computer Science, TELECOM SudParis, UMR 5157 CNRS Samovar, 91011
Evry Cedex, France
{chiraz.trabelsi,sadok.benyahia}@fst.rnu.tn

Abstract. Social Media RSS feeds are the most up-to-date and inclusive releases of information on current events used by the new social media sites such as Twitter and Flickr. Indeed, RSS feeds are considered as a powerful realtime means for real-world events sharing within the social Web. By identifying these events and their associated social media resources, we can greatly improve event browsing and searching. However, a thriving challenge of events identification from such releases is owed to an efficient as well as a timely identification of events. In this paper, we are mainly dealing with event identification from heterogenous social media RSS feeds. In this respect, we introduce a new approach in order to get out these events. The main thrust of the introduced approach stands in achieving a better tradeoff between event identification accuracy and swiftness. Specifically, we adopted the probabilistic Naive Bayes model within the exploitation of stemming and feature selection techniques. Carried out experiments over two real-world datasets emphasize the relevance of our proposal and open many issues.

Keywords: Event Identification, Social Media, RSS feeds, Naive Bayes Model.

1 Context and Motivations

Social media sites such as Twitter and Flickr are using RSS[1] feeds for sharing a wide variety of current and future real-world events. Indeed, RSS feeds are considered as a powerful means of communication for social websites looking to share information on a wide variety of real-world events. These events range from popular, widely known ones, *e.g.,* a concert by a popular music band, to smaller scale local events, *e.g.,* a local social gathering, a protest, or an accident. Social media RSS feeds can typically reflect these events as they happen. By identifying all these events, and their associated social media resources flagged in the RSS feeds items, *e.g.,* photographs, videos, etc., we can enable powerful local event browsing and search, to complement and improve the local search tools that Web search engines provide. In this paper, we address the problem of how to identify, accurately et efficiently, events and their related social media resources over social media RSS feeds items. Actually, social media RSS feeds are defined as a collection of informal data that arrives over time and each RSS feed item is associated

---

[1] Really Simple Syndication.

B. Hong et al. (Eds.): DASFAA Workshops 2013, LNCS 7827, pp. 139–152, 2013.

with some social attributes (features) such as title, description and tags. The content of such RSS feed item, generated from social websites, is particularly useful for real-time identification of real-world events and their associated social media resources, which is the problem that we address in this paper.

Overall, social RSS feeds items, generally exhibit information that is useful for identifying the related events, if any, but this information is far from uniform in quality and might often be fragmented and noisy. Our problem is most similar to the event detection task on Twitter social media [4, 5, 16, 1, 15], whose objective is to identify events in a continuous stream of news documents *e.g.,* earthquakes (Sakaki *et al.,* 2010) or news events (C. Aggarwal and K. Subbian, 2012). However, our problem exhibits some fundamental differences from these event detection approaches that originate in the social RSS feeds resources on which we focus. Specifically, event identification aims to discover and cluster events found in textual news articles. These news articles adhere to certain grammatical, syntactical, and stylistic standards that are appropriate for their venue of publication. Therefore, most state-of-the-art event detection approaches leverage natural language processing tools such as named-entity extraction and part-of-speech tagging to enhance the document representation [4, 5, 1]. In contrast, social RSS feeds resources contain little textual narrative, usually in the form of a short description, title, or keyword tags. Importantly, the lack of both text content and implicite network structure within social RSS feeds resources such as for Twitter messages, renders traditional event identification techniques unsuitable over social RSS feeds resources.

In this paper, we introduce a new approach, called RssE-Miner, for online real-world event identification, that efficiently exploit the Naive Bayes Model for identifying events and their related social RSS feeds resources. The proposed approach consists of three steps: 1) Data preparation and feature selection; 2) Model learning and; 3) Event identification. The main originality of RssE-Miner stands in achieving a meaningful tradeoff between runtime performance and event identification accuracy from social media RSS feeds.

The remainder of the paper is organized as follows. Section 2 thoroughly scrutinizes the related work. We describe later in Section 3 our probabilistic approach for events identification from social media RSS feeds composed of three major steps. An illustrative example is also provides for underpinning the different steps of our approach. The experimental study of our approach is illustrated in Section 4. Section 5 concludes this paper and sketches avenues for future work.

2   Related Work

We describe related work in three areas namely event identification or detection, RSS Feeds studies and Naive Bayes applications.

The event detection task [14, 12] was studied on a continuous stream of news documents with the aim to identify news events and organize them. This is one of the important tasks considered by the topic detection and tracking (TDT) [2]. The topic detection and tracking (TDT) event detection task [14] was studied in a notable collective effort to discover and organize news events in a continuous stream, e.g., newswire, radio broadcast, etc. [16, 1] With an abundance of well-formed text, many of the proposed approaches, *e.g.,* [5] rely on natural language processing techniques to extract linguistically motivated features. Hogenboom et *al.,* [9] proposed an approach for detecting economic events using a semantics-based pipeline. They noted that augmenting documents with semantic terms did not improve performance, and reasoned that inadequate clustering techniques were partially to blame. In our setting, we improve the event identification performance with a judiciously combination of the social media features. Becker et al., in [4], proposed an approach for event detection that aims to partition a set of collected documents from social media into clusters such that each cluster corresponds to all documents that are associated with one event, which is similar to our approach. However, They start from a re-interpretation of exemplar-based SVMs, while we focus more on a probabilistic perspective and Naive Bayes Model. They also propose an additional step for the event detection task during which they use a document similarity metrics to enable online clustering of media to events.

Several efforts have focused on RSS Feeds processing: We can talk about SPEED (Semantics-Based Pipeline for Economic Event Detection) [10] which is a framework that aims at extracting financial events from news articles (announced through RSS Feeds) and annotated with meta-data that makes real-time use possible. It's modeled as a pipeline that reused some of the ANNIE GATE components and develop new ones. We can also refer to SemNews system [11] which is a Semantic Web-based application that aims at accurately extracting information from heterogenous sources. It seeks to discover the meaning of news items. These items are retrieved from RSS Feeds and are processed by the NLP engine OntoSem. Our work also involves the processing of RSS Feeds but using the statistical approach rather than the linguistic approach. The latter constitute a hindrance since linguistic expert uses rules for manually defining event patterns, which is prone to errors.

The literature witness a various applications of the Naive Bayes model. Actually, Naive Bayes showed a powerful performances for automatic categorization of email into folders [6], where email arrives in a stream over time. It was mentioned that Naive Bayes is the fastest algorithm compared to, respectively, MaxEnt, SVM and Winnow. Naive Bayes was also used as a pre-trained model for real-time network traffic classification [13]. Furthermore, Naive Bayes represents, yet, one of the most popular machine learning models applied in the spam filtering domain [17]. Importantly, the learning process of Naive Bayes is extremely fast compared with current discriminative learners, which makes it practical for large real-world applications. Since the training time complexity of Naive Bayes is linear to the number of training data, and the space complexity is also linear in the number of features, it makes Naive Bayes both time and storage efficient for practical systems. This led us to opt for the choice of the Naive Bayes algorithm for social media RSS Feeds processing.

3   Events Identification from Social Media RSS Feeds

As we previously mentioned, identifying social events from social RSS feeds items is a compelling issue for the efficient use of the social RSS feeds releases. Indeed, such events can be useful not only for humans, but also to software agents and applications on the social web. Our hypothesis is that this underlying knowledge can be derived by means of Naive Bayes classifier combined with adequate algorithms for stemming and feature selection. Therefore, the most salient features of our approach, namely RssE-Miner, are as follows: *(i)* It is a domain independent approach, since no domain assumptions are formulated and no predefined knowledge is needed and; *(ii)* It relies on the Naive Bayes model for identifying the RSS feed items related to the same event.

Problem definition: Given a set of social RSS feed resources associated with events, the problem that we address in this paper is how to identify the events that are reflected in the social RSS feed resources, and to correctly assign the resources that correspond to each event. We cast our problem as a classification problem over social RSS feed resources, *e.g.,* photographs, where each social RSS feed resource includes a variety of "features" with information about the resource. Some of these features, *e.g.,* title, description, tags, are manually provided by users, while other features, *e.g.,* upload or con- tent creation time, are automatically generated by the social website. As the formal definition of "event", we adopt the version used for the Topic Detection and Tracking (TDT) event detection task over broadcast news [18].

Definition 1. (EVENT) *An event is something that occurs in a certain place at a certain time.*

In what follows, we introduce the whole process of the proposed RssE-Miner approach. It consists of three steps: 1) Data preparation and feature selection; 2) Model learning and; 3) Event identification. At a glance, Figure 1 provides a visual representation of the RssE-Miner approach.

3.1   Data Preparation and Feature Selection

While social RSS feeds resources present challenges for event detection, they also exhibit opportunities not found in traditional news articles. Specifically, social RSS feeds resources usually have a wealth of associated features such as, *e.g.,* title, description, tags, as well as automatically generated information, *e.g.,* content creation time. Individual feature might be noisy or unreliable, but collectively they provide revealing information about each social RSS feed resource, and this information is valuable to address our problem of focus. There are many design choices for feature representation. In this paper, we made use of the conventional bag of words representation of text items. Hence, each RSS feed item, is represented as a bag of words $\{w_1, w_2, \ldots, w_k\}$. In this respect, common refinements techniques such as discarding common stop-words, pronouns, articles, prepositions and conjunctions are applied. A stemming algorithm provided by Lucene[2] is also used for reducing morphologically similar words, *e.g.,* "starting", "starts" and "start", to the root word, *i.e.,* "start".

Thereafter, we proceed to the feature selection stage in order to identify the most salient features for our model learning. For this purpose, we made use the CFS (Correlation based Feature Selection) algorithm proposed by Hall et *al.,* in [8, 7]. Actually,

---

[2] An open-source search engine that provides full text indexing and searching capabilities, http://lucene.apache.org/

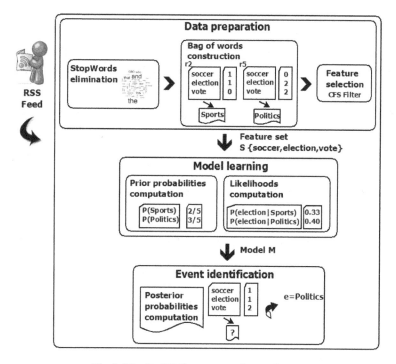

Fig. 1. The RssE-Miner approach at a glance

CFS is a simple filter algorithm[3] that ranks feature subsets according to a correlation based heuristic evaluation function. The bias of the evaluation function is toward subsets that contain features that are highly correlated with the class and uncorrelated with each other. Irrelevant features are ignored because they will have low correlation with the class. While, redundant features are screened out as they will be highly correlated with one or more of the remaining features. CFS algorithm is based on a heuristic evaluation function defined as follows :

$$M_S = \frac{k\overline{h_{cf}}}{\sqrt{k + k(k-1)\overline{h_{ff}}}}. \tag{1}$$

where M_S is the heuristic "merit" of a feature subset S containing k features, $\overline{h_{cf}}$ is the mean feature-class correlation ($f \in S$), and $\overline{h_{ff}}$ is the average feature-feature intercorrelation. The numerator of Equation (1) can be thought of as providing an indication of how predictive of the class a set of features are; the denominator of how much redundancy there is among the features. In this paper, Best first heuristic search is used [8].

---

[3] The time complexity of CFS is quite low. It requires $m((n^2 - n)/2)$ operations for computing the pairwise feature correlation matrix, where m is the number of instances and n is the initial number of features.

3.2   Model Learning

The second step of RssE-Miner aims to determine the likelihood of a RSS feed resource r_i for which the target event is e ($e \in$ a set of events $\mathcal{E}$), given the bag of word $\mathcal{W}$ representing r_i. The variable $\mathcal{W}$ represents a bag of words $\{w_1, w_2, \ldots, w_k\}$ that are extracted from a RSS feed r_i. Hence, a RSS feed resource (represented by $\mathcal{W}$) is identified to be related to an event e_i if the following expression holds:

$$\forall j \neq i \quad \frac{P(\mathcal{W}|e_i)}{P(\mathcal{W}|\bar{e}_i)} > \frac{P(\mathcal{W}|e_j)}{P(\mathcal{W}|\bar{e}_j)} \tag{2}$$

Where $P(\mathcal{W}|e)$ represents the likelihood that $\mathcal{W} = \{w_1, w_2, \ldots, w_K\}$ is produced given the event is e, and $P(\mathcal{W}|\bar{e})$ represents the likelihood that $\mathcal{W}$ is produced given the event is not e. In the Naive Bayes approach, statistical independence is assumed between each of the individual words in $\mathcal{W}$. Under this assumption, the likelihood of $\mathcal{W}$ given an event e is approximated as:

$$P(\mathcal{W}|e) = \prod_{i=1}^{K} P(w_i|e). \tag{3}$$

This expression can be alternatively be represented with a counting interpretation as follows:

$$P(\mathcal{W}|e) = \prod_{i=1}^{K} P(w_i|e)^{C_{w|\mathcal{W}}}. \tag{4}$$

Where $\mathcal{V}$ represents the vocabulary set of words in all training examples. In this interpretation, the occurrence count $C_{w|\mathcal{W}}$ within $\mathcal{W}$ of each word w in the vocabulary set $\mathcal{V}$, is used to exponentially scale the score contribution of that word.

The likelihood function $P(\mathcal{W}|\bar{e})$ is generated as follows:

$$P(\mathcal{W}|\bar{e}) = \frac{1}{K_{\mathcal{E}} - 1} \sum_{\forall e_i \neq e} P(\mathcal{W}|e_i) \tag{5}$$

Where $K_{\mathcal{E}}$ represents the total number of known events. This expression assumes a uniform prior distribution over all events.

In practice the likelihood function $P(w|e)$ is estimated from training examples using maximum *a posteriori* probability (MAP) estimation with Laplace smoothing as follows:

$$P(w|e) = \frac{N_{w|e} + N_V P(w)}{N_{\mathcal{W}|e} + N_V} \tag{6}$$

In this Equation, N_V is the total number of words in the vocabulary used in all training examples, $N_{w|e}$ is the number of times that word w occurs in the training examples related to the event e, and $N_{\mathcal{W}|e}$ is the total number of words in the training examples related to the event e. The term $P(w)$ represents the prior likelihood of word w occurring independently of the event. This likelihood function is also determined using MAP estimation with Laplace smoothing as follows:

$$P(w) = \frac{N_w + 1}{N_W + N_V} \tag{7}$$

In this Equation, N_w is the number of occurrences of the specific word w in the training set examples and N_W is the total count of all words from the N_V word vocabulary in the training set examples.

3.3 Event Identification

Once the model learning step is performed, we proceed with testing the model for identifying the appropriate event. Classification will occurs when event probability is computed. Indeed, the main goal of the event identification step is to find the class label e ($e \in$ a set of events $\mathcal{E}$) which is most likely to generate the RSS feed resource r. Hence, given that the RSS feed resource r is represented as a bag of words $\{w_1, w_2, \ldots, w_k\}$, RssE-Miner tries to assign the most probable class label (event $e \in \mathcal{E}$) to the RSS feed resource based on these words. Therefore, the class label e for r is the most likely event, given the known words of r, is defined as:

$$e = arg \max_{e \in \mathcal{E}} P(e|r) = arg \max_e P(e)P(r|e). \tag{8}$$

Hence, according to Equation 3 and Equation 8, we have:

$$e = arg \max_{e \in \mathcal{E}} P(e) \prod_i P(w_i|e). \tag{9}$$

The event e which have the best score according to Equation 9 is therefore selected.

Example 1. Suppose that once performing the Data preparation and feature selection step, we obtain the model learning example depicted by Table 1. And let us assume that the Laplace smoothing is equal to 1. Then, the prior probabilities for each class label and the probabilities for every word are computed as following:

Table 1. Model learning example

w	soccer	election	vote	Event label
r1	1	0	0	Sports
r2	1	1	0	Sports
r3	0	0	1	Politics
r4	0	1	1	Politics
r5	0	2	2	Politics

Table 2. P(*word*|*Event label*) computation

word	Event label	P(*word*|*Event label*)
election	Sports	0.33
election	Politics	0.40
soccer	Sports	0.50
soccer	Politics	0.10
vote	Sports	0.17
vote	Politics	0.50

$$P(Sports) = 2/5.$$

$$P(Politics) = 3/5.$$

The word "election" occurs 1 times in "Sports" resources.

The total number of words in "Sports" resources = 1+1+1= 3. Then, we have:

$$P(election|Sports) = (1+1)/(3+3) = 1/3.$$

The word "election" occurs 3 times in "Politics" resources.
    Then
$$P(election|Politics) = (3+1)/(7+3) = 2/5.$$

Table 2 resumes this stuff.

We proceed by tackling the same example mentioned above for event identification task. Hence, we provide in Table 3 an example of a resource for which we are looking for the associated event.

Table 3. A RSS feed resource example

w	soccer	election	vote	Event label
r6	1	1	2	?

- e_j=Sports :

$$P(r6|Sports) = P(soccer|Sports)P(vote|Sports)^2 P(election|Sports)$$
$$= 0.0048.$$

- e_j=Politics :

$$P(r6|Politics) = P(soccer|Politics)P(vote|Politics)^2 P(election|Politics)$$
$$= \mathbf{0.010}.$$

Then the event with the highest posterior probability, is selected.

$$e_j = Politics.$$

Fig. 1 (page 4) clearly illustrates the different steps of the RssE-Miner approach for treating the aforementioned example.

4   Experimental Results

To evaluate the performances of our approach for events identification from social media RSS feeds, we carried out experiments on two real world datasets collected from the online photo management and sharing application Flickr[4]. In order to analyze the accuracy of our approach we adopted the common evaluation measures, namely *Accuracy*[5], *Precision*[6] and *Recall*[7] [3]. We describe, in what follows, the datasets characteristics and the baseline models used for evaluating our approach. Thereafter, we present the results from our experiments.

---

[4] http://www.flickr.com/
[5] Ratio of correctly classified instances to the total number of instances in the test set.
[6] The proportion of RSS feeds resources correctly associated to the event e from those associated to e.
[7] The proportion of resources correctly associated to the event e from those actually related to e.

4.1 Datasets Description

We collected our dataset from Flickr using the Flickr API[8]. It consists of RSS Feeds of two real world datasets namely Upcoming and Last.fm.

 - **Last.fm (dense dataset)** : It consists of all Flickr photographs that were manually tagged by users with an id corresponding to an event from the Last.fm music event catalog[9]. The Last.fm dataset contains 3356 images spread over 316 unique events. The Last.fm dataset is considered to be dense since it includes only events in the area of music. It is more likely to find many resources per event.
 - **Upcoming (sparse dataset)** : It consists of all photographs that were manually tagged by users with an event id corresponding to an event from the Upcoming event database[10]. These Upcoming tags provide the "ground truth" for our classification experiments. Each photograph corresponds to a single event, and each event is self-contained and independent of other events in the dataset. The Upcoming dataset contains 5778 images spread over 362 unique events. The Upcoming dataset is considered as a sparse dataset since it contains different kind of events which are of public interest. It is less likely to find two or more items that belong to the same event. Indeed, it includes fewer items per event.

4.2 Training Methodology

We train our approach for event identification on data from the Upcoming dataset, and test them on unseen Upcoming data, as well as Last.fm data. We order the photographs in the Upcoming dataset according to their upload time, and then divide them into two equal parts. We use them as training and test sets. Hence, we use the training set to train classifiers for the classification task. The second datasets of the Upcoming data and the Last.fm data are used as test sets, on which we report our results. We chose a time-based split since it best emulates real-world scenarios, where we only have access to past data with which we can train models to classify future data.

4.3 Baseline Models

To the best of our knowledge, domain independent event identification from social RSS feeds have never been modeled before. Thus, for enhancing the effectiveness of our approach, we have selected three baselines models as follows:

 - **Most Popular Events Identified:** For each event, we counted in how many resources it occurs and used the resources ranked by event occurrence count. For each event, the resources are randomly selected.
 - **Most Popular Event Aware Extracted:** Events are weighted by their co-occurrence with a given event. Then, resources are ordered without validation.

---

[8] http://www.flickr.com/services/api/
[9] http://www.last.fm/events
[10] http://www.upcoming.org

- **SMO:** We compare our approach vs the Becker et al.(2010) [4] approach (we only make use of its classification-based technique part). In fact, Becker et al.(2010) used SVM as a classifier with similarity scores as features to predict whether a pair of documents belongs to the same event. They selected Weka's sequential minimal optimization implementation. In this respect, We implement Naive Bayes as a part of the Weka[11] software system.

4.4   Efficiency of Our Approach

We report, in the following, the obtained results averaged over 10 test runs. We empirically decide to use only description, title and tags features. Indeed, the presence of other features such as location is an indication of document dissimilarity.

Accuracy: Figure 2 (Left), depicts averages of accuracy on Upcoming dataset. Figure 2 (Right), depicts averages of accuracy on Last.fm dataset. On both datasets, SMO shows the highest accuracies. In fact, SMO outperforms RssE-Miner - by notable 1% in the case of Upcoming dataset and by notable 1% in case of Last.fm dataset, but the difference is not statistically significant. However, the performance of RssE-Miner could likely be improved by applying a more sophisticated smoothing method than Laplace. RssE-Miner accuracy, is acceptable since we are looking for the best tradeoff between runtime performance and classification accuracy.

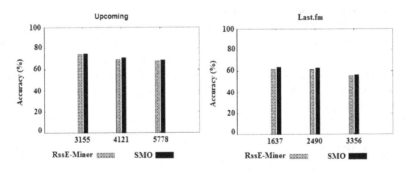

Fig. 2. Left: Averages of *Accuracy* on Upcoming dataset; **Right**: Averages of *Accuracy* on Last.fm dataset

Running Time: As mentioned above, our main goal is to achieve a meaningful trade-off between runtime performance and classification accuracy. Table 4 depicts that our approach has succeeded in fulfilling this task. Thus, according to these results, we can point out that our approach outperforms SMO approach. In fact, as expected, the Runtime of the SMO approach are much slower than those achieved by our approach for both datasets. We note that RssE-Miner is by far the fastest approach. It takes no more than 2.278(s) on Upcoming dataset and no more than 0.739(s) on Last.fm dataset. In

---

[11] http://www.cs.waikato.ac.nz/ml/weka/

particular, our approach outperforms the SMO approach by a large and statistically significant margin. To this end, RssE-Miner makes real-time use possible. This is of paramount importance in the case of event identification in social media RSS Feeds as faster processing of data enables one to make better informed decisions.

In all, evaluation underlines fast and accurate performance by applying our approach. Indeed, achieved results show that event identification using Naive Bayes model can work in near real-time without obvious decrease in accuracy.

Table 4. Average of runtime on the Upcoming and Last.fm datasets

	Instances	Features	Events	Runtime(s)	
				RssE-Miner	SMO
	3155	30	203	**0.596**	34.213
Upcoming	4121	33	280	**1.215**	66.392
	5778	36	362	**2.278**	115.33
	1637	21	171	**0.18**	24.571
Last.fm	2490	27	243	**0.519**	50.143
	3356	23	316	**0.739**	85.829

4.5 Effectiveness of Our Approach

We present in Fig. 3 and Fig. 4- precision as well as recall measures on Upcoming and Last.fm datasets. Indeed, according to the sketched histograms, we can point out that our approach outperforms both baselines. On Upcoming dataset, the average recall achieves high percentage for higher value of N, *i.e.,* the number of extracted events. Indeed, for N = 58, the average Recall is equal to 0.742, showing a drop of 98.38% compared to the average Recall for N = 36. On Last.fm dataset, the average recall achieves high percentage for higher value of N. Indeed, for N = 46, the average Recall is equal to 0.878, showing a drop of 99.4% compared to the average Recall for N = 41. In this case, for a higher value of N, by matching resources with their corresponding events, the proposed approach can achieve event identification task successfully. In addition, the average precision of RssE-Miner outperforms the two baselines. On Upcoming dataset, our approach achieves the best results when the value of N is around 58. In fact, for N = 58, it has an average of 68.3% showing an exceeding about 4% against the first baseline and around 59.6% against the second one. On Last.fm dataset, our approach achieves the best results when the value of N is around 41. In fact, for N = 41, it has an average of 87.3% showing an exceeding about 12.9% against the first baseline and around 64.5% against the second one. These results highlight that the proposed approach can better improve event identification task even for a high number of extracted events.

4.6 Online Evaluation

We present in Fig. 5 the runtime of RssE-Miner. Since it is hard to measure the exact runtime of the proposed approach, we simulated an online execution of our approach

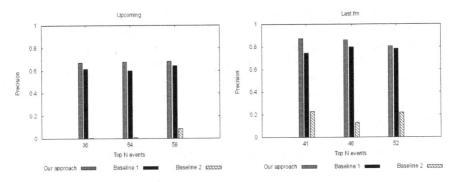

Fig. 3. Left: Averages of *Precision* on Upcoming dataset; **Right**: Averages of *Precision* on Last.fm dataset

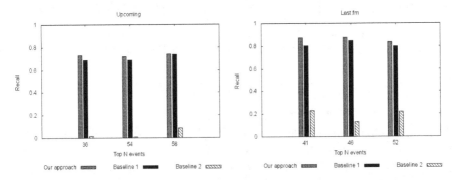

Fig. 4. Left: Averages of *Recall* on Upcoming dataset; **Right**: Averages of *Recall* on Last.fm dataset

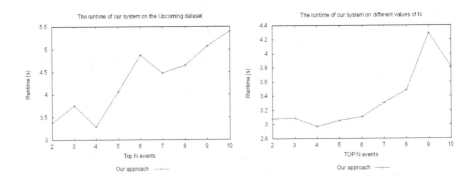

Fig. 5. Left: The runtime of our system online with different values of N on the Upcoming dataset; **Right**: The runtime of our system online with different values of N on the Last.fm dataset

among the Upcoming as well as the Last.fm datasets with different values of N, *i.e.,* the number of extracted events, ranging from 2 to 10. Hence, for each Flickr RSS Feed, we report the average runtime of the related top N events extracted. With respect to Fig. 5, the maximum value of runtime is about 5.403(s) in the Upcoming dataset and about 4.292(s) in the Last.fm dataset, whereas the minimum value is around 3.292(s) in the Upcoming dataset and about 2.975(s) in the Last.fm dataset which is efficient and satisfiable.

5   Conclusion and Future Work

In this paper, we have tackled the challenge of event identification in social media RSS Feeds. We have formulated this task as a real-time problem and introduced a novel probabilistic approach for events mining from heterogenous social media RSS Feeds, called RssE-Miner, in order to get out these events. In particular, our approach relies on a better tradeoff between event mining accuracy and swiftness by applying the probabilistic Naive Bayes model to Flickr data. Our experiments suggest that our approach yields better performance than the baselines on which we build. To the best of our knowledge, event identification (using Naive Bayesian model) in such social media RSS Feeds have never been modeled before. In future work, we will focus on further study other more sophisticated smoothing method than Laplace to improve Naive Bayes performance. Our future research will focus also on event ontology enrichment. Indeed, from these events, we aim to enrich an event ontology. Such an ontology is useful in providing accurate, up-to-date information in response to user queries.

References

1. Aggarwal, C., Subbian, K.: Event detection in social streams. In: Proceedings of the 12th SIAM International Conference on Data Mining, SDM 2012, pp. 624–635. SIAM/Omnipress, Anaheim (2012)
2. Allan, J. (ed.): Topic Detection and Tracking: Event-Based Information Organization. Kluwer Academic Publishers (2002)
3. Baeza-Yates, R., Berthier, R.N.: Modern Information Retrieval. Addison-Wesley Longman Publishing Co., Inc., Boston (1999)
4. Becker, H., Naaman, M., Gravano, L.: Learning similarity metrics for event identification in social media. In: Proceedings of the 3rd ACM International Conference on Web Search and Data Mining, WSDM 2010, pp. 291–300. ACM (2010)
5. Becker, H., Naaman, M., Gravano, L.: Beyond trending topics: Real-world event identification on twitter. In: Proceedings of the 5th International Conference on Weblogs and Social Media, ICWSM 2011. The AAAI Press, Barcelona (2011)
6. Bekkerman, R., Mccallum, A., Huang, G.: Automatic categorization of email into folders:benchmark experiments on enron and sri corpora. Technical Report IR-418, University of Massachusetts, Amherst, USA (2004)
7. Hall, M., Frank, E., Holmes, G., Pfahringer, B., Reutemann, P., Witten, I.: The weka data mining software: an update. SIGKDD Explorations 11, 10–18 (2009)
8. Hall, M.A.: Correlation-based Feature Selection for Machine Learning. Doctoral thesis, The University of Waikato (April 1999)

9. Hogenboom, A., Hogenboom, F., Frasincar, F., Kaymak, U., van der Meer, O., Schouten, K.: Detecting economic events using a semantics-based pipeline. In: Hameurlain, A., Liddle, S.W., Schewe, K.-D., Zhou, X. (eds.) DEXA 2011, Part I. LNCS, vol. 6860, pp. 440–447. Springer, Heidelberg (2011)

10. Hogenboom, F., Hogenboom, A., Frasincar, F., Kaymak, U., van der Meer, O., Schouten, K., Vandic, D.: SPEED: A semantics-based pipeline for economic event detection. In: Parsons, J., Saeki, M., Shoval, P., Woo, C., Wand, Y. (eds.) ER 2010. LNCS, vol. 6412, pp. 452–457. Springer, Heidelberg (2010)

11. Java, A., Finin, T., Nirenburg, S.: Semnews: A semantic news framework. In: Proceedings of the 21st National Conference on Artificial Intelligence, pp. 1939–1940. AAAI Press (2006)

12. Kumaran, G., Allan, J.: Text classification and named entities for new event detection. In: Proceedings of the 27th Annual International ACM SIGIR Conference on Research and Development in Information Retrieval, SIGIR 2004, pp. 25–29 (2004)

13. Li, R.D.W., Abdin, K., Moore, A.: Approaching real-time network traffic classification. Technical Report RR-06-12, Department of Computer Science, Queen Mary, University of London, London (2006)

14. Papka, R., Allan, J., Lavrenko, V.: On-line new event detection and tracking. In: Proceedings of the 21st Annual International ACM SIGIR Conference on Research and Development in Information Retrieval, SIGIR 1998, pp. 37–45. ACM (1998)

15. Reuter, T., Cimiano, P., Drumond, L., Buza, K., Schmidt-Thieme, L.: Scalable event-based clustering of social media via record linkage techniques. In: Proceedings of the 5th International Conference on Weblogs and Social Media, ICWSM 2011. The AAAI Press, Barcelona (2011)

16. Sakaki, T., Okazaki, M., Matsuo, Y.: Earthquake shakes twitter users: real-time event detection by social sensors. In: Proceedings of the 19th International Conference on World Wide Web, WWW 2010, pp. 851–860. ACM, New York (2010)

17. Song, Y., Kolcz, A., Giles, C.L.: Better naive bayes classification for high-precision spam detection. Softw. Pract. Exper. 39(11), 1003–1024 (2009)

18. Yang, Y., Pierce, T., Carbonell, J.: A study on retrospective and on-line event detection. In: Proceedings of the 21st ACM International Conference on Research and Development in Information Retrieval, SIGIR 1998, pp. 28–36. ACM, New York (1998)

A Method of Pre-detecting Privacy Leak in Social Network Service Using Collaborative Filtering

Chungha Kim, Kangsoo Jung, and Seog Park

Department of Computer Science, Sogang University, Seoul, Korea
krieiter@gmail.com, {azure84,spark}@sogang.ac.kr

Abstract. Nowadays, the number of SNS user is increasing very quickly and importance of information in SNS is getting higher. While other traditional web services treat only photos and articles of users, SNS covers more sort of information. SNS privacy issue receives more attention because SNS users upload and update information about them in real time spontaneously. Existing researches mainly concentrate about extracting profiles from each user or connections between users. In this paper, we present a possible privacy attacking method for an attacker may use, and show that the method can be adopted in real social network service form as the third party application. Especially, we show how the application works to analyze threats to which SNS users can be exposed, and to show in what way users should react when using SNS.

Keywords: Privacy, Social Network Service, Collaborative Filtering, Data Privacy.

1 Introduction

SNS(Social Network Service)is an online service that people share their news and communicate each other. Proliferation of smartphone and development of wireless network bring an explosion of SNS. Users share various private information such as Daily life, Image, Video, Bookmark, Game, Location information through SNS. This kind of information is valuable for company because it is really helpful to analyze people's preference for marketing. That is why they try to find a way to analyze SNS information.

Collaborative filtering is one of the methods to extract user's preference or find user who have specific preference. It is based on assumption that user who behaves similarly has similar preferences. In this assumption, if user X and user Y behave similarly, they help each other to predict their future preference consequently although they do not cooperate explicitly. It is effective method to predict user's unexposed preference, however it can make a privacy issue because user's preference can be inferred without agreement. In this paper, we focus on possible privacy leakage that is occurred by third party application developer or SNS provider using collaborative filtering.

The rest of the paper is organized as follows. Section 2 surveys and analyzes related works. Section 3 explains proposed method to attack user's privacy by collaborative

B. Hong et al. (Eds.): DASFAA Workshops 2013, LNCS 7827, pp. 153–163, 2013.
© Springer-Verlag Berlin Heidelberg 2013

filtering. Section 4 presents the results of the experiments by implementation of third party application in the Facebook. In section 5, we summarize the contributions of the proposed technique and discuss our future research works.

2   Related Works

According to increasing of the number of SNS user, privacy becomes a significant issue because users upload their private information in SNS. There are many researches about privacy in SNS. These researches are classified into three categories. (1) User's profile extraction[6,7,8] (2) Access control for SNS[5] (3) Network topology analysis in SNS[2,3,4]. Our work focuses on problem of user's profile extraction in SNS. In this section, we introduce about user's profile extraction and show the privacy attack method.

2.1   User's Profile Extraction Method

Social network is modeled as a graph model [3]. In graph model, each vertex represents user, and each edge represents relationship between users. Each vertex and edge might have additional information such as user's profile and preference. In this section, we explain existing attack method to expose individual's information in SNS. These techniques are classified into identifying relationship between user's and exposing user's sensitive information.

2.1.1 Identifying Relationship between User's in SNS

In [8], they propose an attack method to identify relationship among users in SNS. Using this method, it traces each user's connection with other users and identifies entire network's connection among users. Generally, user allows other user who has relationship in SNS to share their information. An attacker bribes other user to get connection information between users and tries to infer whole network's connection information using this information.

2.1.2 Exposing User's Sensitive Information in SNS .

Jha[7] suggests an automatic profiling method and framework for SNS. Jha's method is based on an insight that a lot of SNS use email address as an ID, and most users use the same email address for SNS. Attacker queries to SNS using user's email address list. SNS provider returns a confirmation about existence of user who uses that email address. Attacker aggregates user's public profile in SNS like this way, and integrates user's profile to infer extra information about user. This method collects user's public profile, but sensitive information that user do not want to expose might be inferred by integration of this profile.

2.2   Motivation

Existing researches focus on extracting user's profile or identifying relationship between users using public information. That is why traditional techniques concentrate to get whole social network information rather than target user's information. In this paper, we consider privacy leakage that can be occurred by inference of user's profile that user do not expose.

3   Privacy Leakage Detection Method Using Collaborative Filtering

3.1   Problem Description

We model social network as a graph model G = (V, E). As we explained previously, each vertex represents user and edge represents user's relationship. SNS user can add specific item(e.g. Last.fm) or click like button to express their preference(e.g. Facebook),   and this information is collected by SNS provider. SNS provider broadcasts this information to other user to promote building relationship among users. We define this item that user expresses their preference as 'Interesting item'. We show this interesting item and relationship information can be used as information to infer user's unexposed preference.

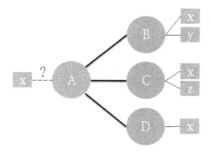

Fig. 1. Example of privacy leakage

[Fig. 1] is a part of whole social network. An attacker does not know relationship between item x and user A. user B, C, D is user A's friend and they have their own interesting item that is exposed in public. In this situation, attacker assumes that user A also has relationship between item x as interesting item, although this relationship is not exposed. Because all other user B, C, D own item x as interesting item. This assumption is based on an insight that SNS user makes a relationship with other user who has similar preference.

That is why we can infer specific user's interesting item although it is not exposed using related user's interesting item information. We define this kind of privacy leakage as a privacy leakage using collaborative filtering.

For example, if most user who has relation with some user expose sensitive information such as sexual taste or religious beliefs, some user can not hide his sensitive information because attacker can infer user's preference by collaborative filtering. We prove how to do this kind of privacy leakage is occurred by implementation of third party application in Facebook. These kind of privacy leakage can not be solved by existing privacy policy because it is based on inference.

3.2   Privacy Leakage Detection Algorithm

In this work, we assume that attacker get permissions to access target user's profile and related user's profile and interesting item. We describe proposed algorithm that how to infer target user's interesting item using related user's profile and interesting item list.

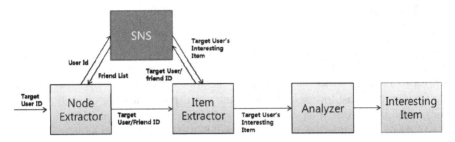

Fig. 2. Process of proposed algorithm

[Fig. 2] is a process of proposed technique. Node extractor is a component to extract target user's related user list. Item extractor extracts related user's interesting item. And analyzer infers target user's interesting item using Item extractor's results. At first, attacker inputs target user's Id to node extractor. Node extractor extracts target user's related user list and pass this list to Item extractor. Item extractor extracts related user's interesting item and pass it to Analyzer. Analyzer infers target user's interesting item using sensitivity estimation function. Pseudo algorithm is as follows.

Table 1. Pseudo algorithm of proposed algorithm

```
Input: target user's ID, sensitivity threshold
Output: target user's interest item list S
Item's sensitiv N = {}
Related user list U_adj = {}
Target user's interest item list T_u = {}
Related user's interest item list T_adj = {}
Similarity between target user and related user M_u = {}
U_adj = read_adjacent_users(u)
items T_u = read_related_items(u)

    foreach U_adj^i in U_adj do
        T_adj = T_adj read_related_items(U_adj^i)
        M_u^i = similarity(read_related_items(U_adj^i), T_u)
    done
    foreach T_adj^i in T_adj do
        sensitivity N_i = sensitivity(T_u, T_adj, M_u, T_adj^i)
    done
    S = T_adj
```

```
foreach N_i in N do
   if N_i <
      then S = S - T_adj^i
   endif
done
S = S - T_u
return S
```

3.3 Sensitivity Estimation Function

Proposed technique estimated sensitivity of interesting item that is extracted by Item Extractor to detect privacy leakage degree. Sensitivity estimation function uses target user and related user's interesting item list as an input to calculate sensitive item list. tf × idf method[11] is applied to calculate item's sensitivity and Jaccard Coefficient[12] is used to calculate similarity among interest items.

We define that target user's related user's list as T_u , related user's interesting item list as T_{adj}, similarity between target user and related user as S_u, and item I as a input value. Sensitivity function s(T_u, T_{adj}, S_u, I)is described as follow. This function's output is a sensitivity of target user's interesting item I.

$$\text{sentivitiy function } s\big(T_u, T_{adj}, S_u, I\big) = \sum_{I \in T_{adj}^i} S_u^i \times \frac{\max(popularity(I))}{popularity(I)} \qquad (1)$$

Function popularity(x) represents a proportion of interest item x in whole network. The reason why we calculate popularity(x) is to normalize each item's sensitivity. That is, each item's sensitivity is decided by popularity in whole network.

S_{uv} is a function that calculates similarity degree between target user and related user's interesting item. t_u and t_v are item set and similarity equations is as follows.

$$S_{uv} = \text{similarity}(u, v) = \frac{|t_u \cup t_v| + 1}{|t_u \cup t_v| - |t_u \cap t_v| + 1} \qquad (2)$$

|x| is a number of list x's item. We apply Jaccard Coefficient[12] method to calculate similarity. However, we use a multiplicative inverse of Jaccard Coefficient because Jaccard Coefficient represents a difference between two items.

3.4 Example

We explain proposed algorithm using example. [Fig.3] is an part of social network.

A,B,C,D represent user and user B, C, D is user A's friend and they have their own interesting item that is exposed in public. User A's interesting items are j,k and user B's are j,k,l, C's are k,l,m and D's are l,m,n. In this case, each item's popularity is as follows.

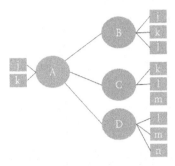

Fig. 3. Example of Social network

Table 2. Popularity of Item

Item	j	k	l	m	n
popularity	3	4	12	4	8

In this example, l is the highest popularity, That is why max(popularity(t)is 12. Proposed technique calculate privacy sensitivity degree by relationship between user A's items and user B,C,D's items.

For doing this, we calculate User B,C,D's similarity with User A. Each User's similarity is calculated using each item's similarity[Eqn.1].

$$similarity(A, B) = \frac{|\{j, k, l\}| + 1}{|\{j, k, l\}| - |\{j, k\}| + 1} = \frac{3 + 1}{3 - 2 + 1} = \frac{4}{2} = 2.0$$

$$similarity(A, C) = \frac{|\{j, k, l, m\}| + 1}{|\{j, k, l, m\}| - |\{k\}| + 1} = \frac{4 + 1}{4 - 1 + 1} = \frac{5}{4} = 1.25$$

$$similarity(A, D) = \frac{|\{j, k, l, m, n\}| + 1}{|\{j, k, l, m, n\}| - |\emptyset| + 1} = \frac{5 + 1}{5 - 0 + 1} = \frac{5}{5} = 1.0$$

And then, we calculate each item's sensitivity using similarity. Item i and m's sensitivity is like this[Eqn. 2].

$$sensitivity(A, l) = (2.0 + 1.25 + 1.2) \times \frac{12}{12} = 4.45$$

$$sensitivity(A, l) = (2.0 + 1.25 + 1.2) \times \frac{12}{12} = 7.35$$

As a result, we assume that User A can have relationship with m and m is the most sensitive item because m has the highest sensitivity value except j, k that has direct relationship with User A.

4   Experiment

In this section, we implement proposed technique as a third party application in Facebook and show how to expose user's sensitivity information without user's agreement.

4.1   Experiment Design

For this experiment, we actually implemented the Facebook third party application that is shown in [Fig. 4].

Fig. 4. Facebook application for experiment

We collected necessary data from spontaneous people who have actually used Facebook and let them answer about sensitivity degree of extracted data. Because it is so difficult to measure how users feel threatened by privacy leakage except by survey. We implemented experiment application as the actual third party application to get actual user's answer and compare this result with survey results.

A existing technique as a comparison group used FilmTrust[16], a cooperative filtering method that expects the grade of movie with using reliability between users and linked information in SNS. The third party application implemented for the survey collects necessary data from voluntary users, and let them choose the integer from 1 to 5 as the value of sensitivity about the exposed top 10 items among extracted items. A questionnaire is as follows.

1) How does this item describe you well?
2) Would you mind if your relationship with this item is exposed to the public?

4.2   Experiment Result

We analyzed degree of privacy threats that users sensed, and compared with existing technique through the experiment we performed. Also, we collected necessary data from 16 users for 11 days and additionally carried out the survey of the same users again for 3 days to ask whether to make public the item.

4.2.1   Sensitivity of Each Item

With the experiment we performed in this section, we analyzed the degree of privacy threats users sensed, compared with existing technique, and let users answer directly to how they are sensitive to each item. The result is shown in [Fig. 5].

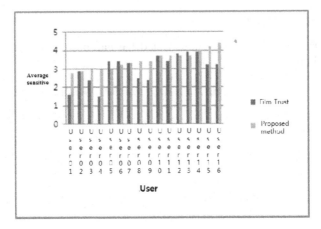

Fig. 5. Sensitivity of each item

[Fig. 5] includes average of the grade users answered about extracted items at the vertical axis. The higher the grades are, the more extracted items are sensitive in terms of privacy. We used RMSE(Root Mean Square Error) to compare difference between the extracted items with using each technique and sensitive items users believed actually. In general, RMSE is a technique, which estimates performance of the cooperative filtering system. It is used to compare with errors between predictive values and real values of techniques. Therefore, it is a kind of checking system for false-positive. The RMSE values of the proposed and existing technique is shown in Table 3 below.

Table 3. RMSE value

	RMSE FilmTrust	Proposed Technique
RMSE	1.96875 (95% confidence interval [1.310, 1.764])	1.5375 (95% confidence interval [1.610, 2.327])

To calculate the value of RMSE,   we assume that the value of sensitivity is 5 equal to the maximal value of sensitivity for the 10 items extracted by each algorithm. And we defined error as difference between this value and values that users answered actually.

4.2.2   Number of Sensitive Item

As a result of this survey, users doesn't feel sensitive in the case of the items that has 1, 2 as the value of the sensitivity. They do not seem to be related much with this technique's goal to show that extracting sensitive items is possible through the information extracted in SNS. Therefore, in this paper, we analyzed the items that has 3, 4, 5 as the value of sensitivity among items that users answered. The result is shown in [Fig. 6].

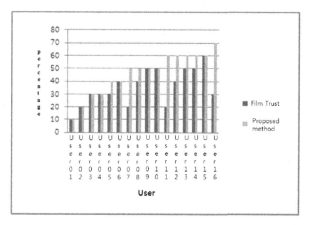

Fig. 6. Sensitive item rate

[Fig. 6] includes the ratio of items that has 3, 4, 5 as the value of the sensitivity and are exposed after being extracted through each technique, according to users, at the vertical axis. In an experimental group, with the proposed technique, the number of items that has 3, 4, 5 increases or is not changed rather than existing technique. Averagely, confidence interval from 10.6% to 96%: [0.3811, 1.7439], It found out more sensitive items.

4.2.3   Actual Sensitivity

In addition to the result of the survey on the sensitivity of each items, we made the second question more directly in the questionnaire to analyze more accurately whether users do not want to make public these items actually. We ask users to answer whether they want to make public these about the same items. The response is shown in [Fig. 7].

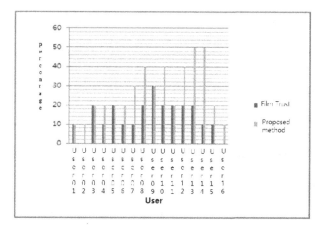

Fig. 7. Actual Sensitive item Rate

Averagely, users said they do not want about 26.875% items to be made public. Compared with 14.375% derived by the previous technique, this figure is advanced. Therefore, we know that this technique can find out more the items that users do not want to make public than the previous technique.

Through this result, we show that suggested technique is actually able to find out related items that user does not make public and that privacy leakage can occur actually by adjacent nodes.

5   Conclusion

Previous researches on attack and detection techniques in SNS focused simply on user's public profile and relationship among users. On the other hand, suggested technique detects each user's privacy threats that can be hidden in SNS under the insightful fact that network among users has more meaning than simple topology.

In this research, we implement actually the technique as a form of the third party application in SNS, from these threats. Also, according to collecting data on actual users and the result of the survey, we prove that privacy leakage can be caused by this attack method.

We proved the privacy threats that can occur actually when user has an access to the third party applications. Hence, This result might be very important. The contribution points in this research are as follows.

1. We show the possibility of private information leakage that user doesn't want to make it public
2. We prove effectiveness and possibility of attack method using adjacent user's information
3. We verify range and level of privacy threats of the actual third party applications

Acknowledgement. ddfThis research work is supported by DGISH Global CPS Center.

References

1. Su, X., Khoshgoftaar, T.M.: A survey of collaborative filtering techniques. Advances in Artificial Intelligence, 1–19 (2009)
2. Narayanan, A., Shmatikov, V.: De-anonymizing social networks. In: 2009 30th IEEE Symposium on Security and Privacy, pp. 173–187 (2009)
3. Diaz, C., Troncoso, C., Serjantov, A.: On the impact of social network profiling on anonymity. In: Borisov, N., Goldberg, I. (eds.) PETS 2008. LNCS, vol. 5134, pp. 44–62. Springer, Heidelberg (2008)
4. Backstrom, L., Dwork, C., Kleinberg, J.: Wherefore Art Thou R3579X? Anonymized social networks, hidden patterns, and structural steganography. In: WWW (2007)
5. Beato, F., Kohlweiss, M., Wouters, K.: Enforcing access control in social networks. In: Proc. HotPets (2009)

6. Kamahara, J., Asakawa, T., Shimojo, S., Miyahara, H.: A community-based recommendation system to reveal unexpected interests. In: Proceedings of the 11th International Multimedia Modelling Conference, Washington, DC, USA, pp. 433–438 (2005)
7. Balduzzi, M., Platzer, C., Holz, T., Kirda, E., Balzarotti, D., Kruegel, C.: Abusing social networks for automated user profiling. In: Jha, S., Sommer, R., Kreibich, C. (eds.) RAID 2010. LNCS, vol. 6307, pp. 422–441. Springer, Heidelberg (2010)
8. Korolova, A., Motwani, R., Nabar, S., Xu, Y.: Link privacy in social networks. In: Proceeding of the 17th ACM Conference on Information and Knowledge Mining (2008)
9. Facebook, Facebook Markup Language (FBML) (July 4, 2012),
   http://developers.facebook.com/docs/reference/fbml/
   (July 11, 2012)
10. Calandrino, J.A., Kilzer, A., Narayanan, A., Felten, E.W., Shmatikov, V.: You might also like: privacy risks of collaborative filtering. In: IEEE Symposium on Security and Privacy, pp. 231–246 (2011)
11. Salton, G., McGill, M.J.: Introduction to modern information retrieval. McGraw-Hill (1986)
12. Tan, P.-N., Steinbach, M., Kumar, V., Grover, R., Vriens, M., Grover, R.E., Vriens, M.E.: Introduction to Data Mining. Journal of School Psychology 19(1), 51–56 (2005)
13. Kamahara, J., Asakawa, T., Shimojo, S., Miyahara, H.: A community-based recommendation system to reveal unexpected interests. In: Proceedings of the 11th International Multimedia Modelling Conference, Washington, DC, USA, pp. 433–438 (2005)
14. Baeza-Yates, R., Ribeiro-Neto, B.: Modern Information Retrieval, pp. 28–30. Addison Wesley (1999)
15. Golbeck, J., Hendler, J.: FilmTrust: movie recommendations using trust in web-based social networks. In: Proceedings of 3rd IEEE Consumer Communications and Networking Conference, vol. 1, pp. 282–286 (2006)
16. Zhou, B., Pei, J.: Preserving Privacy in Social Networks Against Neighborhood Attacks. In: 2008 IEEE 24th International Conference on Data Engineering, pp. 506–515 (2008)

The Strategies for Supporting Query Specialization and Query Generalization in Social Tagging Systems[*]

Jia-Ling Koh, Kuang-Ting Chiang, and I-Chih Chiu

Department of Information Science and Computer Engineering,
National Taiwan Normal University, Taipei, Taiwan, R.O.C.
jlkoh@csie.ntnu.edu.tw

Abstract. In this paper, we design a tag ranking method to provide multi-level keyword suggestion. The suggested keywords are used to effectively filter query results, which helps users to perform query specialization in social tagging systems. Besides, error-tolerant set containment queries are used to support various degrees of query generalization. We propose an index structure, which aggregates similar tag sets into clusters. A bounding mechanism is provided to efficiently deal with query processing for error-tolerant set containment queries on tag sets. These strategies can be used to support generalizations of a query. A systematic performance study is performed to show the effectiveness and the efficiency of the proposed methods.

Keywords: query keyword suggestion, index structure, query processing, error-tolerant set containment query, social tagging system.

1 Introduction

Social tagging has become widely popular in many social sharing systems. Social tagging systems enable users to freely label web resources of interest with tags for describing the contents or semantics of the resources, which provides a simple solution for annotations. In recent studies, it has been shown that tags can be used to improve searching, navigation and recommendations effectively [3].

In social tagging systems, keyword based search is a popular way for discovering required data of interest from a huge collection of sharing resources. The effectiveness of data retrieval mainly depends on whether the given queries properly describe the information needs of users. However, it is not easy to give a precise query because most queries are short (less than two words on average) and many query words are ambiguous [13]. Using a general keyword with board semantics as a query usually causes a huge amount of data returned. It is difficult for users to explore and find objects which satisfy their search needs from a long list of returned results [2]. Consider a query "apple" on Flickr as an example. Because various semantics are represented by the word "apple", the query results may contain the images of apples,

---

[*] This work was partially supported by the R.O.C. N.S.C. under Contract No. 101-2221-E-003-025.

B. Hong et al. (Eds.): DASFAA Workshops 2013, LNCS 7827, pp. 164–178, 2013.

the products of Apple company, the New York City etc., which are mixed together in the returned ranked list. It is costly for users to sequentially browse the objects in the ranked list in order to find a favorite image of red apples. For solving this problem, some search systems perform clustering on the query results, which separates the results into groups of similar objects. However, the effectiveness of the clustering results usually depends on the numbers of clusters. It is still a challenge to dynamically decide the number of clusters for properly clustering the query results. Moreover, the computational cost of performing clustering on the query results is usually high.

In order to provide convenient and efficient exploring for the returned query results, some search engines organize the results according to some facet attributes such as year or data type. However, most of the systems define the used facets in advance. For the results with tags, a tag can be viewed a binary facet attribute of objects. In spirit to recent works on automatic facets generation [6], we would like to dynamically suggest keywords from the tags of the returned query results. Users can select one of the suggested keywords to be inserted into the original query in order to form a more specific query for effectively narrowing the query results. For example, "fruit", "mac", and "NewYork City" are good keywords suggested to further distinguish different information needs of the query "apple". Adding one of the suggested keywords into the original query could provide query reformulation for *specialization* in order to navigate part of the original query results.

Moreover, a challenge on querying social tagging systems is that an object might not be well tagged completely. It is possible that an object satisfies the need of users but is not in the query results because the tag set of the object does not contain all keywords in the query. For example, suppose an object o is annotated by a set of tags {"apple", "Manhattan", "Central Park"}. Although object o contains the concept "New York City", it is not annotated the association tag explicitly. Then "New York City" is a missing tag of object o. When a query {"apple", "New York City", "Manhattan"} is given, the object o is lost from the query results if a query result is required to contain all the tags in the query. For getting object o in the query results, the user may like to ask a query more general than the previous one, which is obtained by removing one of the keywords from the original query. However, it is not easy to decide which keyword should be removed from the original query. We define the *Error-Tolerant Set Containment* query (ETSC query for short) on tag sets in order to support query reformulation for *generalization*. Given a query q consisting of keywords and a query threshold θ, the goal of an ETSC query is to find all the objects whose tag sets contain at least $|q| \times \theta$ tags in the query. In the previous example, if the query threshold is 2/3, the objects containing at least two keywords in the query are returned as results of the ETSC query. Accordingly, ETSC queries can support various degrees of query generalization by controlling the query threshold. How to provide efficient processing of ETSC queries is another issue studied in this paper.

Our work is motivated from the challenges mentioned above. The contributions of this paper are as follows:

(1) We apply tag ranking methods to level-wisely suggest query keywords for providing specializations of a query. Besides, we apply an ETSC query to provide generalizations of a query.

(2) We propose an index structure, which aggregates tag sets with similar members into clusters. A bounding mechanism is proposed to efficiently deal with query processing for ETSC queries.
(3) A systematic performance study is performed to verify the effectiveness and the efficiency of the proposed methods.

The rest of this paper is organized as follows. The related works are discussed in Section 2. Section 3 introduces the proposed method of suggesting keywords for query specialization. The index structure for supporting error-tolerant containment queries is introduced in Section 4. The results of performance evaluation on the provided methods are reported in Section 5. Finally, Section 6 concludes the works proposed in this paper.

2   Related Works

Faceted search offers a more flexibility to explore the results by specifying various facet refinements, which has become influential over the last few years [10]. That is why faceted search has become increasingly popular in online information access systems. The facets in typical faceted search systems for navigation are usually static, which often don't change with different queries. Accordingly, recent works began to study how to dynamically discover a small set of facets and values from a large set of search results which are most interesting to a user or most relevant to a query. The Facetedpedia system proposed in [7] was a faceted retrieval system, which was designed for information discovery and exploration in Wikipedia. The system can automatically and dynamically construct facet hierarchy for navigating the set of Wikipedia articles resulting from a keyword query. [14] proposed a system named TEXplorer, which integrated keyword search of documents with the aggregation and exploration functionalities on the corresponding structured data of documents. A measurement was defined for effectively ranking dimensions of structured data in order to facilitate user exploration on relevant documents. In spirit to the success of faceted search interface, we would like to suggest keywords as facets for effectively navigating query results in a tagging system.

In [11], a tag clustering approach was proposed to solve the syntactic and semantic tag variations during searching and browsing activities of users. The normalized Levenshtein similarity is used to compute the similarity between each pair of tags for discovering syntactic clusters of tags. Then the cosine similarity based on co-occurrence vectors is measured to find clusters with semantics related tags. The discovered clusters of tags are used for searching and browsing the tag spaces. However, the computation cost of finding both syntactic and semantic tag clusters is very high.

Tag cloud is a visual form to display the representative tags for a collection of objects, which is usually applied to facilitate browsing and searching process of the objects. In [8], experiments were performed to study in what situations using a tag cloud is better than doing search. The results showed that the participants preferred tag clouds if the information-seeking task was more general. In that situation, tag clouds make the users get better familiar with the data and further provided more focused queries. [7] studied how the various features used to construct a tag cloud, such as font size and word placement, affect the various tasks of searching, browsing, impression formation and recognition. The typical approaches usually select the most popular tags as tag clouds to summarize a collection of resources. In [12], the author considered that it lacks metrics to evaluate the tag selection results. Accordingly, a set of various metrics for evaluating tag clouds, such as coverage, overlap, cohesiveness, relevance, popularity, and independence, were provided. Besides, four algorithms of tag selection were provided, each of which was designed for optimizing a single metric function of tag clouds. Similarly, in [9], it was studied that whether frequency is the most suitable criterion for tag ranking. A set of functions were provided to perform tag ranking, which includes the frequency scoring, TF-IDF scoring, graph-based scoring, and diversity or novelty of the members of the tag cloud. The proposed ranking methods were used to find one level suggestion of tag clouds. In this paper, we apply the ranking strategies of tags to construct multi-level tag suggestions, which can provide specializations of query results just like facet query interface.

3   Multi-level Keyword Suggestion for Query Specialization

3.1   Terms Definition

Let WDB denote a database of web resources, which contains a set of objects: $WDB = \{o_1, ..., o_n\}$. We assume that each object o_i in WDB has a set of tags, denoted by $o_i.tagset$. A query q in our system consists of a nonempty set of keywords. The returned query results of q from WDB, denoted O_q, is the set of objects whose tag sets are the supersets of q, i.e., $O_q = \{o \mid o \in WDB \wedge q \subseteq o.tagset\}$. The set of candidate keywords for specializing query q, denoted CT_q, is defined to be the union of the tag sets of O_q except the tags in q, i.e., $CT_q = \{t \mid t \in o.tagset \wedge o \in O_q\} - q$. In other words, we will suggest keywords for query specialization from the tags which are contained in the tag sets of the query results.

3.2   Tag Scoring Functions

We consider the suggested keywords for query specialization should have the following two properties: 1) a suggested keyword is semantically related to the query keyword, and 2) the semantic concept represented by a suggested keyword is different from the concepts of other suggested keywords. Let ST_q denote the set of tags selected into the suggested list, which is an empty set initially. We design the scoring function for evaluating the suggestion score of a tag t_i in CT_q by a weighted sum of two parts: the value of r_score function and the value of d_score function, as follows:

$$S\_score(q, t_i) = w \times r\_score(q, t_i) + (1 - w) \times d\_score(t_i, ST_q). \tag{1}$$

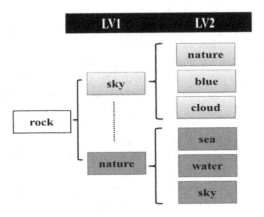

Fig. 1. An example of multi-level query keyword suggestion

For a tag t_1, if another tag t_2 appears in most of the objects which t_1 appears, it usually shows that t_2 is semantically related to t_1. Accordingly, we use the normalized confidence value of an association rule $q \rightarrow t_i$ to be the related score of t_i for q. The $r\_score(q, t_i)$ is defined as follow:

$$conf(q, t_i) = \frac{f(t_i \mid O_q)}{\mid O_q \mid} . \tag{2}$$

$$r\_score(q, t_i) = \frac{conf(q, t_i) - \min_{t \in CT_q}(conf(q, t))}{\max_{t \in CT_q}(conf(q, t)) - \min_{t \in CT_q}(conf(q, t))} . \tag{3}$$

The $d\_score(t_i, ST_q)$ scoring function should return higher value for the tags which are less semantically related to the tags in ST_q. Therefore, the $d\_score(t_i, ST_q)$ is defined to be the average of dissimilarity between t_i and each tag in ST_q as follow:

$$d\_score(t_i, ST_q) = \frac{1}{\mid ST_q \mid} \sum_{t_j \in ST_q} \frac{1 - sim(t_i, t_j)}{1 + sim(t_i, t_j)} . \tag{4}$$

The $sim(t_i, t_j)$ is evaluated according to the co-occurrence degree of t_i and t_j in O_q as follows:

$$sim(t_i, t_j) = \frac{f(t_i, t_j \mid O_q)}{f(t_i \mid O_q) + f(t_j \mid O_q) - f(t_i, t_j \mid O_q)} , \tag{5}$$

where $f(t_i \mid O_q)$ denotes $\mid\{o \mid o \in O_q \land t_i \in o.tagset\}\mid$ and $f(t_i, t_j \mid O_q)$ denotes $\mid\{o \mid o \in O_q \land t_i \in o.tagset \land t_i \in o.tagset \}\mid$, respectively.

The ranges of both functions $r\_score(q, t_i)$ and $d\_score(q, t_i)$ are between 0 and 1. After testing various setting on w, a better setting of w is 0.3.

3.3   Keyword Suggestion for Query Specialization

According to the scoring function $S\_score(q, t_i)$ defined in the previous section, we adopt an incremental approach to select tags from CT_q to be inserted into ST_q. The following steps are performed to find k_1 suggested query keywords for providing specialized navigation on the query results of q on the first level.

Step 1:   Find O_q from WDB.

Step 2: Compute $CT_q = (\bigcup_{o \in O_q} o.tagset) - q$.

Step 3: $ST_q = \phi$.

    Step 3-1: For each t_i in CT_q, compute $r\_score(q, t_i)$.

    Step 3-2:   For each t_i in CT_q, compute $d\_score(t_i, ST_q)$.

    Step 3-3: Let t_{max} denote the tag in CT_q whose value of $S\_score(q, t_i)$ is the maximum.

        $ST_q = ST_q \cup \{ t_{max} \}; CT_q = CT_q - \{ t_{max} \}$.

    Step 3-4: Repeat Step 3-2 to Step 3-3 until $|ST_q| = k_1$.

Let $ST_q = \{ st_1, st_2, ..., st_{k1} \}$ denote the suggested query keywords on the first level. If users select st_i, it means that users would like to browse the objects whose tag sets contain $q \cup \{ st_i \}$. Accordingly, let $q' = q \cup \{ st_i \}$. The system will find $O_{q'}$ from O_q and compute $CT_{q'}$. For further reducing the query results of q', the above process can be performed recursively. By setting the number of suggested query keywords on the second level, denoted k_2, Step 3 is then performed on $CT_{q'}$ to find k_2 suggested query keywords under st_i on the second-level. An example of two-level keyword suggestions for query "rock" is shown in Fig. 1. The above process can be performed repeatedly to find the suggested keywords for query specialization on the following levels. However, it is not necessarily that providing more suggested query keywords on each level or providing more levels of keyword suggestion is better for navigating the query results. In the experiments, we will observe how the setting of the number of keywords suggested on each level influences the cost of navigating query results.

4   ETSC Query Processing

4.1   Definition of ETSC Queries

Given a set of keywords as a query q, a containment measure function, and a query threshold δ, the goal of an Error-Tolerant Set Containment(ETSC) query is to find the objects in WDB, whose degrees of covering the tags in the query are larger than or equal to δ.

Here, we use *Jaccard containment* [1] to compute the degree of a tag set t_i covering a query q as follows.

$$Jcontain(t_i, q) = |q \cap t_i| / |q| . \tag{6}$$

The value of $Jcontain(t_i, q)$ is between 0 and 1, which means the ratio of tags in q contained in t_i. If $Jcontain(t_i, q)=1$, it indicates that t_i contains all the tags in q.

4.2   Index Structure Construction

In [5], we proposed a multi-level index structure to support efficient similarity searches of tag sets in a social tagging system. Here, we modify our previous work to provide an index structure for processing ETSC queries efficiently.

Each entry in the index structure is composed of a tag set ts and an object list. The object list maintains the identifiers of the objects whose tag sets are equal to ts. Besides, the tag sets of objects in the index structure are organized into a forest. Each tree in the forest contains a set of tag sets with similar elements. The root node of a tree T maintains two features: the result of Union and the result of Intersection on the tag sets stored in T, which are named the outer border of T and the inner border of T, denoted $T.ob$ and $T.ib$, respectively. The tag sets which can be stored in a tree T are controlled by limiting the maximum distance between $T.ob$ and $T.ib$. The distance between two tag sets T_i and T_j is evaluated by the Hamming distance as the following:

$$dist(T_i, T_j) = diff(T_i, T_j) + diff(T_j, T_i), \quad \text{where } diff(T_i, T_j) = |T_i - T_j| \ldots\ldots(7).$$

The threshold $maxd$ is used to limit the maximum distance between the outer and inner borders of a tree in the index structure. Another threshold β is used to limit the maximum number of tag sets stored in a node.

Initially, the index structure is empty. The tag sets of the objects in WDB are inserted into the index structure one by one. Let o_i denote the inserted. If $o_i.tagset$ has already existed in the index structure, the identifier of the object o_i is inserted into the object list of the entry with tag set $o_i.tagset$. Otherwise, it is necessary to find a tree or create a new tree that the $o_i.tagset$ can be inserted into.

The tag set $o_i.tagset$ is allowed to be inserted into the root node of a tree T in the index structure if $T.ib \cap o_i.tagset \neq \phi$ and $dist(T.new\_ob, T.new\_ib) \leq maxd$, where $T.new\_ob = (T.ob \cup o_i.tagset)$ and $T.new\_ib = (T.ib \cap o_i.tagset)$. If there are a number of root nodes of trees that the tag set $o_i.tagset$ can be inserted into, the tree T_{best} which has the smallest difference between its new_ib and new_ob after inserting $o_i.tagset$ is selected. Then the entry for storing $o_i.tagset$ and its object list is inserted into the root node of T_{best} and the borders of T_{best} are updated as follows: $T_{best}.ob = T_{best}.new\_ob$ and $T_{best}.ib = T_{best}.new\_ib$. If such a tree does not exist, a new tree new_T with a root node is created for storing the entry of $o_i.tagset$ and its object list. Besides, both $new\_T.ob$ and $new\_T.ib$ are set to be $o_i.tagset$.

After all the objects are inserted into the root nodes of the trees in the index structure, the splitting process is then performed. If the root node of a tree T contains more than β tag sets, a splitting on T is performed. Because the tag sets stored in T contains all the tags in $T.ib$, for each tag set t_i in T, only the tags in $(t_i - T.ib)$ can be used to further group the similar tag sets in T into sub-trees of T. Accordingly, for each entry with tag set t_i and object list ol_i stored in the root node of T, we set $t_i' = t_i - T.ib$. The tag set t_i' is used to decide which child node the entry of (t_i', ol_i) is assigned to by recursively applying the insertion method performed on the root level. However, the threshold value used to limit the maximum distance between the outer and inner borders of the children nodes of T is updated with a reduced threshold value $maxd'= maxd - |T.ib|$. If such a child node does not exist, a new node is created for storing the entry of (t_i', ol_i), which is then assigned to be a child node of the root node of T.

Assume that *maxd* is set to 4 and β is set to 3. Suppose the tag sets of 8 objects are given as shown in Fig. 2(a), after inserting the tag sets into the index structure, the inner border, the outer border, and the maintained tag sets in each root node are shown as Fig. 2(b). After performing the splitting process, the resultant children nodes are shown as Fig. 2(c).

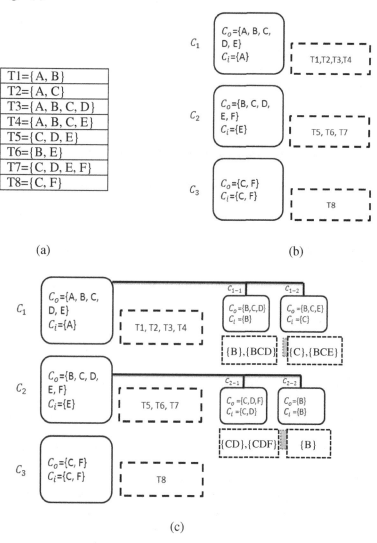

(a)   (b)

(c)

Fig. 2. An example of constructing the index structure

4.3   ETSC Query Processing

When processing an ETSC query for a given query q, the root node of each tree T in the index structure is checked. According to the inner border/outer border of the root node of T, a lower/upper bound of the degree of any tag set in T covering q is estimated, which is denoted by $JC\_LB(q, T)/JC\_UB(q, T)$. The equations used to estimate $JC\_LB(q, T)/JC\_UB(q, T)$ are defined below.

$$JC\_LB(q, T) = |q \cap T.ib|/|q| . \tag{8}$$

$$JC\_UB(q, T) = |q \cap T.ob|/|q| . \tag{9}$$

By comparing $JC\_LB(q, T)$ and $JC\_UB(q, T)$ with the query threshold δ, there are three possible cases.

1) $JC\_UB(t_q, T) < \delta$: all the entries in the root node of tree T can be pruned from the search space.

2) $JC\_LB(t_q, T) \geq \delta$: all the obejcts in the corresponding object lists of the entries in the root node of tree T are retrieved as the results of the ETSC query.

3) $JC\_LB(t_q, T) < \delta \leq JC\_UB(t_q, T)$: If T has subtrees, the checking process on each child node of T should be performed. Otherwise, each entry with tag set t_i and object list ol_i, which is stored in T, should be checked one by one to certify whether $|q \cap t_i|/|q| \geq \delta$. The objects in ol_i are retrieved as the results if t_i satisfies the checking.

Before performing bounding checking on the subtrees of T, the query tag set and query threshold should be modified as follows:

$$q\_new = q - T.ib \quad , \quad \delta_{new} = \frac{(|t_q| \times \delta) - |t_q \cap T.ib|}{|t_q - T.ib|} . \tag{10}$$

Then the same checking mechanism is performed on each subtree of T recursively by using q_new as the query and δ_{new} as the query threshold.

5   Performance Evaluation

5.1   Experiment Setup

A systematic study is performed to evaluate the effectiveness of the suggested keywords for query specialization and the performance efficiency of the proposed index structure for processing ETSC queries. The proposed methods and algorithms were implemented using Java on the Netbean platform. The experiments were performed on a personal computer with Intel Core i5 CPU and 6G megabytes DDR memory and running Microsoft Windows 7.

In the experiments, we used the web image dataset created by NUS's Lab [4]. The dataset includes 269,648 images and the associated tags from Flickr. There are a total of 5,018 unique tags annotated for these images. In data preprocessing, we used the Porter's stemming algorithm to transfer the tags which are plurals to their singular forms.

Table 1. The statistics of frequencies for the tags used as test queries

	High Frequency tags	Middle Frequency tags	Low Frequency tags
Number of queries	64	71	165
Max frequency	20219	2984	990
Min frequency	3001	1001	100
Avg frequency	7101.5	1816.9	468.37

5.2   The Effectiveness of Keyword Suggestion

The goal of keyword suggestion proposed in this paper is to make users find the required objects from the query results more efficiently. Accordingly, we simulated the user navigation model [10] to evaluate the reduced ratio of navigation cost when providing different numbers of suggested tags.

Let c_t denote the cost of deciding whether to select a suggested keyword, c_o denote the cost of browsing an object, and st_i denote the ith suggested keyword. For a query q, when keyword suggestion is not provided, the average cost of finding an object o from O_q is $c_o \times \lceil |O_q|/2 \rceil$. This value is used as a baseline of navigation cost. Let st_i denote the ith suggested keyword. By using the suggested query keyword st_i, the average cost of finding an object in $O_{q \cup \{st_i\}}$ is $c_t \times i + c_o \times \lceil |O_{q \cup \{st_i\}}|/2 \rceil$. For the object o in O_q which does not contain any suggested keyword, the average cost of finding o is $c_t \times k + c_o \times \lceil |O_q - (\cup_{i=1..k} O_{q \cup \{st_i\}})|/2 \rceil$ when k keywords are suggested. The average navigation cost of using suggested keywords is divided by the baseline cost to get the relative cost ratio, where a smaller relative cost ratio indicates a higher reducing of navigation cost. In the following experiments, we assume c_t is two times the value of c_o. Besides, we also evaluate the coverage rate of the suggested keywords. Let $O_c(q)$ denote the set of objects which can be retrieved by selecting any suggested keyword for query specialization on O_q. In other words, $O_c(q) = \cup_{i=1..k} O_{q \cup \{st_i\}}$. The coverage rate of the suggested keywords for O_q is $|O_c(q)|/|O_q|$.

We randomly select 300 distinct tags as test queries, where each query contains a tag. According to the frequencies of tags, we separate the queries into 3 groups: the low frequency tags, the middle frequency tags, and the high frequency tags. The number of queries, maximum frequency, minimum frequency, and average frequencies of each group of queries are shown in Table 1.

The number of suggested keywords on the first level, denoted by k_1, is varied from 5 to 60. The coverage rate on the query results by the suggested k_1 keywords is shown in Fig. 3(a). Besides, the result of relative cost ratio is shown in Fig. 3(b). It is not surprising that providing more suggested keywords gets higher coverage. For the queries with low or middle frequency tags, the average coverage rate achieves 90% when k_1 is 30. For the queries with high frequency tags, when k_1 is 40, the average coverage rate also achieves 90%. It is worth noticing that the increasing of coverage and the decreasing of navigation cost are less significant when k_1 is larger than 30. Besides, for the queries with low frequency tags, the lowest relative cost ratio occurs when k_1 is 20. Accordingly, it indicates that $k_1=20$ is a good choice for suggesting query keywords in the first level.

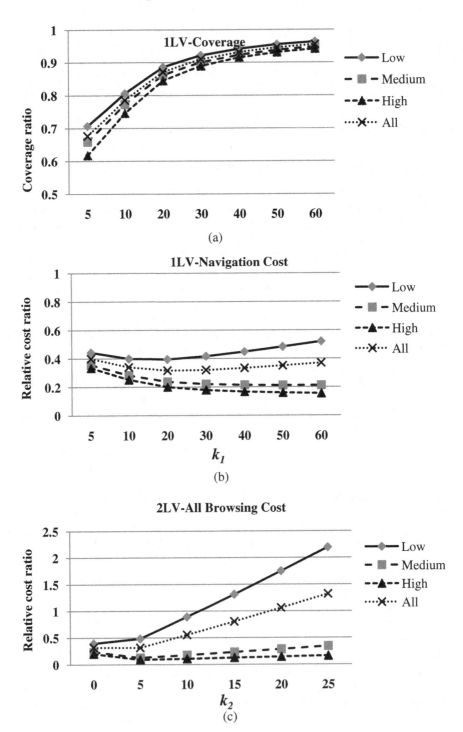

Fig. 3. Performance of suggesting various number of query keywords

In the second experiment, k_1 is fixed to be 20 and the number of suggested keywords on the second level, denoted by k_2 is varied from 5 to 60. As the results shown in Fig. 3(c), for the queries with low frequency tags, the navigation cost keeps increasing when suggesting query keywords on the second level. When k_2 is larger than 15, it even gets higher navigation cost than the cost of the baseline. It indicates that it is not necessary to suggest query keywords on the second level for the queries with low frequency tags. In contrast, for the queries with middle or high frequency tags, $k_2=5$ is a good choice to further reduce the relative cost ratio to below 0.1. The above results show that providing a proper number of suggested query keywords can effectively reduce the navigation cost.

5.3   The Efficiency of ETSC Query Processing

This part of experiments wants to observe how the various setting of parameter *maxd* used in index construction influences the execution time of processing ETSC queries. We also consider the two running-time effects: the number of tags in a query, denoted $|q|$, and the query threshold δ, which may affect the execution time.

1)   Varying the setting of *maxd*

In this experiment, the parameter *maxd* is varied from 200 to 1000. The parameter β is set to be 20. When δ is set to be 1, the execution time of the proposed approach for processing ETSC queries with different query lengths are shown in Fig. 4(a). It shows that a larger values of *maxd* gets shorter execution time for short queries ($|q|=1$ or $|q|=2$). However, it is not necessarily a good choice to set a large value for *maxd* when $|q|=3$ or $|q|=4$. When $|q|$ is fixed to be 4, the execution time for different query thresholds are shown in Fig. 4(b). It shows that a larger *maxd* takes longer execution times when δ is less than 1. Accordingly, the value of *maxd* is set to be a medium value 500 in the following experiments.

2)   Varying the query length $|q|$

The naïve method, which sequentially checks all the tag sets to find the answer, is used as a baseline. By using $\delta=1$ and *maxd*=500, the comparison of the execution time between the proposed method and the baseline by varying $|q|$ is shown in Fig. 4(c).

3)   Varying the query threshold (δ)

In this experiment, *maxd* is set to be 500 and $|q|$ is fixed to be 4. The execution time of the proposed method and the baseline by varying δ is shown in Fig. 4(d).

Fig. 4(c) and 4(d) shows the proposed index structure and query processing strategies improves the efficiency of processing ETSC queries significantly. Besides, various $|q|$ and δ only have slight influence on the execution time of the proposed method.

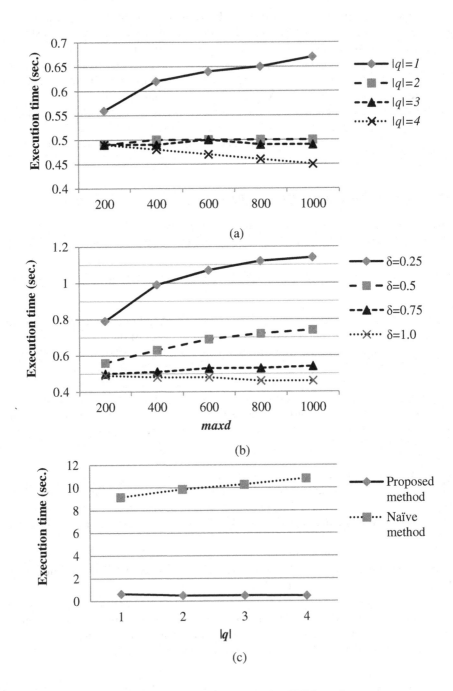

Fig. 4. Execution times of processing ETSC queries

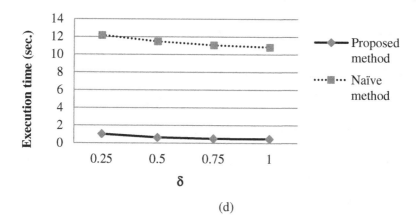

(d)

Fig. 4. (*continued*)

6   Conclusion and Future Work

In this paper, we design a tag ranking method to provide multi-level keyword suggestion. The suggested keywords can be used to effective filtering query results, which
helps users to perform query specialization in social tagging systems. Besides, we
propose an index structure, which aggregates similar tag sets into clusters. A bounding
mechanism is proposed to efficiently deal with query processing for ETSC queries.
These strategies can be used to efficiently provide generalizations of a query. The
results of experiments show the effectiveness of the query keyword suggestion for
query specialization and the efficiency of the proposed ETSC query processing method. How to automatically organize the suggested query keywords into a semantic
concept hierarchy is under our investigation.

References

1. Agrawal, P., Arasu, A., Kaushik, R.: On indexing error-tolerant set containment. In: Proc.
   the ACM International Conference on Management of Data, SIGMOD (2010)
2. Amer-Yahia, S.: I am complex: Cluster me, don't just rank me. In: Proc. the 2nd International Workshop on Business intelligence and the WEB, BEWEB (2011)
3. Carmel, D., Roitman, H., Yom-Tov, E.: Social bookmark weighting for search and recommendation. The VLDB Journal 19, 761–775 (2010)
4. Chua, T.-S., Tang, J., Hong, R., Li, H., Luo, Z., Zheng, Y.-T.: NUS-WIDE: areal-
   worldweb image database from national university of Singapore. In: Proc. ACM International Conference on Image and Video Retrieval (2009)
5. Koh, J.-L., Shongwe, N., Cho, C.-W.: A multi-level hierarchical index structure for supporting efficient similarity search ontagsets. In: Proc. IEEE International Conference on
   Research Challenge in Information Science (2012)

6. Li, C., Yan, N., Roy, S.B., Lisham, L., Das, G.: Facetedpedia: dynamic generation of query-dependentfaceted interfaces for wikipedia. In: Proc. the World Wide Web (WWW), pp. 651–660 (2010)

7. Rivadeneira, A.W., Gruen, D.M., Muller, M.J., Millen, D.R.: Getting our head in the clouds: toward evaluation studies of tagclouds. In: Proc. the SIGCHI Conference on Human Factors in Computing Systems, CHI (2007)

8. Sinclair, J., Cardew-Hall, M.: The folksonomy tag cloud: when is it useful? Journal of Inf. Science 34(1), 15–29 (2008)

9. Skoutas, D., Alrifai, M.: Tag Clouds Revisited. In: Proc. the 20th ACM International Conference on Information and Knowledge Management (ACM CIKM) (2011)

10. Tunkelang, D.: Faceted Search. Synthesis Lectures on Information Concepts, Retrieval, and Services 1(1), 1–80 (2009)

11. Vandic, D., van Dam, J.-W., Hogenboom, F.: A semantic clustering-based approach for searching and browsing tag spaces. In: Proc. of the 26th ACM Symposium on Applied Computing, SAC (2011)

12. Venetis, P., Koutrika, G., Garcia-Molina, H.: On the selection of tags for tag clouds. In: Proc. of the Fourth ACM International Conference on Web Search and Data Mining, WSDM (2011)

13. Wen, J.-R., Nie, J.-Y., Zhang, H.-J.: Query clustering using user logs. ACM Transactions on Information Systems (ACM TOIS) 20(1), 59–81 (2002)

14. Zhao, B., Lin, X., Ding, B., Han, J.: TEXplorer: keyword-based object search and exploration in multidimensional text databases. In: Proc. the ACM International Conference on Information and Knowledge Management (CIKM) (2011)

Extracting Relations among Search Properties Based on the Operational Context of Geographical Information Retrieval Systems

Daisuke Kitayama[1], Junki Matsuo[2], and Kazutoshi Sumiya[2]

[1] Faculty of Infomatics, Kogakuin University
1–24–2 Nishi–shinjuku, Shinjuku, Tokyo 163-8677, Japan
kitayama@cc.kogakuin.ac.jp
[2] Graduate School of Human Science and Environment, University of Hyogo
1–1–12 Shinzaike–honcho, Himeji, Hyogo 670-0092, Japan
nd11g028@stshse.u-hyogo.ac.jp,
sumiya@shse.u-hyogo.ac.jp

Abstract. During searches for suitable objects with geographical information retrieval systems, users can select various search properties such as the region, category, price, distance, and rating. We assumed that there would be implicit connections among the properties in the search operations adopted by users. For example, if a user wants to go a bar, they might search for a bar by changing the rating property to "high" and changing the price property to "moderate." If the user wants to go a bar again, however, they might search for a bar by changing the rating property to "moderate" and changing the price property to "low." Thus, the price property and rating property are connected by the context when searching for the "bar" category. We consider that the connected properties may differ according to the operational context. In this paper, we propose a method for extracting connected properties based on the operation history of user. We also describe an automatic adjustment function based on the connected properties extracted.

Keywords: Automatic Adjustment, Connected Search Properties, Geographical Information Retrieval, Operational Context.

1 Introduction

Information about restaurants and hotels can be retrieved using an online map. For example, people often use Yelp[1], Booking.com[2], Tabelog[3], and HOT PEPPER[4] to plan their travels and determine where they will dine. With these systems, users can select various search properties such as the region, category, price, distance, and rating when searching for suitable places. If a user wants to

---

[1] http://www.yelp.com
[2] http://www.booking.com
[3] http://tabelog.com/
[4] http://www.hotpepper.jp/

B. Hong et al. (Eds.): DASFAA Workshops 2013, LNCS 7827, pp. 179–192, 2013.

search for inexpensive restaurants, they select "low value" in the price property and "restaurant" in the category property. The user is then provided with a list of restaurants that satisfy the selected search properties. We consider that there are implicit connections between the properties of the search operations conducted by users. Thus, if a user changes one property A, they will also change another property B, which is related to property A for that user. For example, if a user changes the price to "higher price," they may also change the rating to "higher rating." Therefore, we consider that the user wants to search for an expensive restaurant so that the price property and the rating property are connected.

The user's requirements may different in specific search contexts. Therefore, the connected properties may depend on the user's search context. If the user has connected properties for the price and rating in the search context "bar," they might also have connected properties for the price and distance in the search context "restaurant" because they may prefer not to use their car when consuming alcoholic drinks. Thus, the connected properties are changed in different user search contexts. Therefore, we need to extract the connected properties in each search context.

In this study, we propose a method for extracting relationships among search properties depending on the operational context based on association rule mining. Using the extracted relationships, we develop an automatic adjustment function for search properties to reduce the number of user operations.

The remainder of this paper is organized as follows. We describe the concept of connected properties based on the user's operation history and other work related to this topic in Section 2. In Section 3, we explain how we extract the relationships among connected properties. We present and evaluate our experiments in Section 4. Finally, we discuss the use of our method in Section 5.

2   Automatic Property Adjustment Based on Operational Context

2.1   Concepts of Operational Context and Connected Properties

We consider that a search property may be connected with another search property during geographical information search. For example, if a user wants to go a bar, they might search for a bar by changing the rating property to "high" and changing the price property to "moderate." If the user wants to go a bar again, they may search for a bar by changing the rating property to "moderate" and changing the price property to "low." In this case, the price property and the rating property are connected by the context when searching for a "bar" category. In general, the price property is not always connected to the rating property. However, we consider that the connection may change depending on the specific context. If the user wants to visit a hot dog shop, they may change the price property and the distance property because they may want to eat a hot dog soon. Thus, we considered that extracting connections among the search properties in

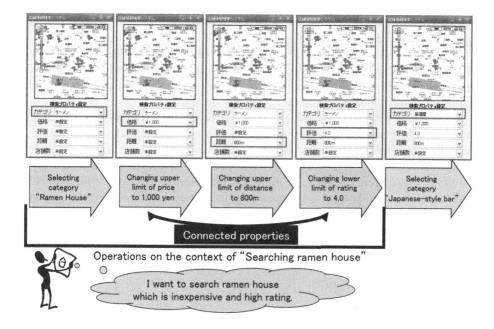

Fig. 1. Concept of connected properties depending on operational context

a user's operation history would allow us to adjust the properties automatically when a user changes a search property, depending on its connection with another property. This could allow users to obtain geographical information with fewer operations.

We explain the concept of a connected property that depends on the operational context using Figure 1. In this figure, the user selects the category "ramen house" and changes the price, distance, and rating, before changing the category. We consider that the operations performed after selecting "ramen house" depend on the context when "searching for a ramen house." We define the operational context as the leading operation of the property in a specific context. In this case, the operational context is the category and the value is "ramen house." The operational context is the search property with the highest priority according to the user. Therefore, we determine the operational sequence as the flow from changing the property until changing the same property as the context, and we use the leading operation of the property as the operational context. In a specific context, the operational context is not changed by user. We consider that a user changes the properties by trial and error in an operational context.

The frequency of user operations could reduce the connected properties extracted. Thus, we extract the connected properties by association rule mining. We treat the operational context as the antecedent and a pair of operations in the search context as the consequence. The association rule determines combinations with strong relationship when a specific phenomenon occurs. Using this rule, we can extract the connected search properties that appear frequently in a specific search context.

The connected properties may have positive or negative correlations. A positive correlation indicates that the value of a property increases with the value of another property; a negative correlation indicates that the value of a property increases when the value of another property decreases.

We describe a typical scenario as follows. A user searches frequently for "Italian food" and "bar." The user wants to go to a restaurant with a good rating, regardless of whether the restaurant is expensive. Initially, our system presents "Italian food," "bar," and other categories as the choices. If the user selects "Italian food," the system presents the rating properties such as "high," "moderate," and "low," because the rating property is often used in the context of searching for "Italian food." If the user selects the "high" rating property, the system changes the price property to "high" implicitly because the rating property and the price property are connected in this user's search context. In this way, our system reduces the user's operations by exploiting connected properties.

2.2   Related Works

Previous studies have analyzed the relationships among multiple properties in search systems. Kobayashi et al. [1] proposed an automatic adjustment method for multiple properties that displayed a constant number of search results based on user's operation history via an online map and a category tree. We enhanced their approach to consider other properties such as the price, rating, and distance, as well as map operations and category selection.

Cho et al. [2] proposed an interface for inputting multiple properties by drawing a trapezoid. The trapezoid included two properties, i.e., the rating on the vertical axis and the distance on the horizontal axis. Users could input multiple properties intuitively via this interface. In this method, users had to select multiple properties manually and consider the relationship among these properties. In contrast, our method can extract relationships among multiple properties, so users only need to perform general operations.

Bostandjiev et al. [3] proposed a music recommendation interface based on automatic property adjustment using social media such as twitter and Facebook. The automatic property adjustment method was similar to our approach, but they used connected properties as static relationships from social media, whereas we use connected properties from the user's operation history as dynamic relationships.

Dalal et al. [4] generate comments to rank news articles using Hodge decomposition for multiple aspects such as the user rating and the quality of comments. This ranking was effective as an evaluation standard for the user. However, their method only generated a static ranking for news articles. In geographic information search tasks, the user preference often changes depending on the user's search context; hence, our approach needs to adjust multiple properties dynamically.

Liu et al. [5] proposed a recommendation method based on user clustering to extract preferences for each property. However, the user's property requirements changed depending on the user's search contexts during geographical information search tasks. In general, similar user preferences are important during geographical information search tasks. Therefore, these methods may be complementary.

Next, we discuss previous studies on geographical information search based on the user context. Biancalana et al. [6] proposed a recommendation system based on context extraction such as "working" and "traveling" in the present location and time. In this system, the context was the user's environment. However, we use the search context to refer to the user's current search properties such as "I want to search in the "Italian food" category."

Baltrunas et al. [7] proposed a geographical information recommendation system based on the relationship between the user context and categories. In their system, the user inputs context explicitly. In contrast, our method automatically extracts the user's search context by association rule mining.

3   Extracting Relations among Search Properties

Our proposed system has a normal mode and a question mode. Users can conduct general restaurant searches in the normal mode. We store the user's operation history in this mode to extract the relations among search properties. However, users can select the options displayed by the system in the question mode. The question mode generates the property options based on the relations extracted from among the search properties.

In this section, we described how we extract the relations among the search properties from the user's stored operation history. We use association rule mining to extract the relations. Initially, we divide the user's operation history and we treat the divided operation history as the substrate for association rule mining. Next, we describe how we calculate the support and confidence when extracting relations.

3.1   Dividing User's Operation History

We divide the user's operation history into separate ranges for each operational context. We considered that links between two operations with the same property indicate trial and error when the user is setting a primary property. For example, the user changes the category property to "sushi bar" and the price property to "lower price," before changing the rating to "higher points." This series of operations is considered to be a specific behavior in the search context of "sushi bar." After these operations, the user changes the category property to "Italian food." We consider that there is no relationship among "Italian food," "lower price," and "higher points." Thus, we consider that the operations conducted after property X is changed to property value A, and until property X is changed to property value B, are related to property value A. Therefore, we divide the user's operation history by changing the same property and the end of a search

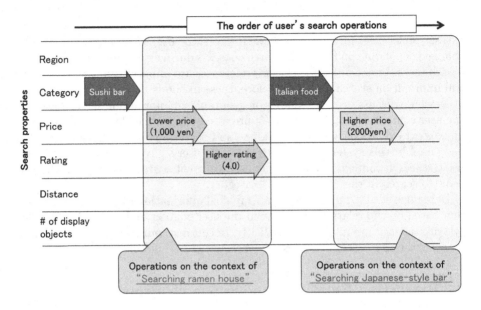

Fig. 2. Example of division of the user's operational history

session. The search contexts are not only the category property, because there are also other properties. We use all of these properties to divide the operational history. Therefore, the divided operation histories overlap with each other.

Figure 2 shows an example of the division of the user's operational history. The vertical axis shows the number of property operations in Figure 2. Initially, the user selects the category "sushi bar" (operation 1), before changing the price to "lower" (operation 2) and the rating to "higher" (operation 3). The user then selects the category "Italian food" (operation 4) and changes the price to "higher" (operation 5). In this case, we consider that operations 2 and 3 share the same operational context as operation 1, that operation 5 has the same operational context as operation 4, and that operations 3 and 4 share the same operational context as operation 2.

We divide the user's operation history as described previously to generate the substrate for association rule mining.

3.2 Extracting Relations Using Association Rule Mining

We extract association rules as the relations among search properties using the divided operation history. The association rule identifies combination phenomena with strong simultaneity and high probability that when a phenomena occurs, another phenomena will occur. Using this rule, we determine the connection among search properties that appear frequently in a specific search context. We use the support measure and the confidence measure to calculate the strength of a relationship.

Table 1. List of search properties

Property name	Property value selection
Region	Using Google Maps
Category	Selecting from list box
Price	Selecting upper limit from combo box
Rating	Selecting lower limit from combo box
Distance	Selecting upper limit from combo box
Number of objects displayed	Selecting upper limit from combo box

$$\text{support}\,(X \Rightarrow Y) = \frac{\text{count}\,(X \cup Y)}{|D|} \tag{1}$$

$$\text{confidence}\,(X \Rightarrow Y) = \frac{\text{count}\,(X \cup Y)}{\text{count}\,(X)} \tag{2}$$

where D are the divided operation histories; X denotes an operational context, which is a property value such as "sushi bar" or "low price"; and Y is a pair of properties and its operation, such as changing the price to "lower price" or changing the distance to "wider area." The support measure is the ratio of $X \cup Y$ to D. Thus, this measure indicates the popularity in the user's operation history. The confidence measure is the ratio of $X \cup Y$ to the segment of the divided operation histories that includes X. Thus, this measure indicates the popularity in the user's operational context.

If a rule has a high support value and a high confidence value, an operation that contains $X \cup Y$ appears frequently, so the operational context X, a pair of properties, and its operation Y have a strong relationship. Therefore, we multiply support by the confidence to obtain a score that indicates the relationship strength.

4   Experiment

4.1   Prototype System

We developed a prototype system to evaluate our method. Our system has two modes. In the normal mode, users can conduct general operations such as restaurant searches with yelp. This mode stores the user's property operations to extract relationship. Figure 3 shows the normal mode interface. We use six properties (see Table 1). This interface comprises the online map component, category selection component, component for changing properties, and component for showing detailed information. The online map component uses the Google

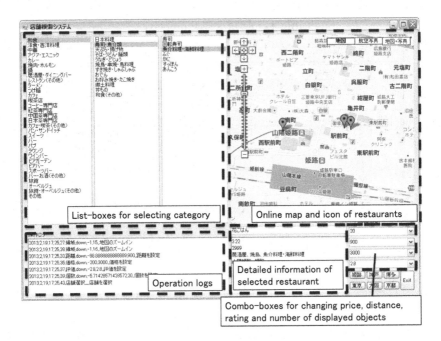

Fig. 3. Normal mode interface

Maps API[5]. This component shows the objects retrieved as icons. Users can change the search region and select objects using the map component. The category selection component has three list boxes and the component for changing the properties has four combo boxes. Users can select one choice from each of these list boxes and combo boxes. The component for showing detailed information provides the details of the restaurant selected via the map component. We use Tabelog[6] to obtain restaurant information. We collect this data via the Web API of Tabelog.

4.2 Experimental Setting

There were three participants in this study who were all university students. The participants performed 10 search tasks at random in five areas (Himeji, Osaka, Kobe, Tokyo, and Hakata in Japan) and the 12 situations shown in Table 2. These areas included those that were familiar and unfamiliar area to the participants.

We assumed that the participant's search contexts were different during each task. When each participant finished a task, we asked the participants to report their search objectives using a questionnaire based on free-form answers to obtain the experimental results.

---

[5] https://developers.google.com/maps/

[6] http://tabelog.com/

Table 2. Search tasks used in the experiment

Task 1	Searching for a restaurant for a New Year's party with many friends.
Task 2	Searching for a restaurant to kill time until a meeting at the station.
Task 3	Searching for a restaurant for lunch or dinner with children.
Task 4	Searching for a restaurant for dinner with an old friend.
Task 5	Searching for a restaurant at a previously arranged business trip destination.
Task 6	Searching for a restaurant for working outside the office.
Task 7	Searching for a restaurant to meet a client.
Task 8	Searching for a restaurant to take rest during a date in one's hometown.
Task 9	Searching for a restaurant for lunch or dinner with a few close friends.
Task 10	Searching for a restaurant to have a drink after work.
Task 11	Searching for a restaurant to read a book in a relaxed mood
Task 12	Searching for a restaurant to have lunch with the family during a car trip.

Table 3. Experimental results: the connected properties extracted

	A	B	C
Rank 1	Great distance $\Rightarrow$ Rating & Price (P)	Medium distance $\Rightarrow$ Rating & Price (P)	High rating $\Rightarrow$ Distance & Region (P)
Rank 2	Medium region $\Rightarrow$ Rating & Price (P)	Japanese food category $\Rightarrow$ Rating & Distance (P)	Medium distance $\Rightarrow$ Rating & Price (P)
Rank 3	Cafe category $\Rightarrow$ Price & Region (P)	Inexpensive $\Rightarrow$ Rating &Distance (P)	Medium price $\Rightarrow$ Distance & Region (P)
Rank 4	Japanese-style bar category $\Rightarrow$ Rating & Price (P)	Medium region $\Rightarrow$ Price & Distance (P)	Inexpensive $\Rightarrow$ Distance & Region (P)
Rank 5	Ramen house category $\Rightarrow$ Rating & Price (P)	Medium region $\Rightarrow$ Rating & Price (P)	High rating $\Rightarrow$ Price & Distance (P)

The total number of operations was 626. The number of operations when changing properties was 268 and the number of operations when selecting restaurants was 358. We obtained 255 divided user operation histories. The divided user operation histories included very short segments and those that comprised one or two operations. Therefore, we used 110 divided user operation histories to extract the connected properties.

4.3   Experimental Results

Table 3 shows the results for the connected properties, which were extracted based on the operational context. This table shows the top five ranked scores based on the association rule. The left side of $\Rightarrow$ shows the operational context, and the right side of $\Rightarrow$ shows the connected properties. P indicates a positive correlation between connected properties, whereas N indicates a negative correlation between connected properties (we could not extract higher ranked connected properties with negative correlations).

We extracted different operational contexts and connected properties for each participant. We describe each result in detail, as follows. Participant A usually moved by car and wanted to save money, but also wanted to go to a restaurant with a high rating. For ranks 1 and 2, the extracted connected properties were consistent with the participant's intentions. We can see that participant A connected the rating and the price, even when the search area was wide. For ranks 4 and 5, the extracted connected properties were the rating and the price in the operational contexts of the categories "ramen house" and "Japanese-style bar." For rank 3, however, the extracted connected properties were the region and price in the operational context of the category "cafe." Participant A often used the "cafe" category as a meeting point, and searched for "cafe" in tasks 2, 5, 6, and 11. We consider that Participant A wanted to search for a "cafe" that was located nearby. We confirmed that the connected properties differed according to the operational context.

Participant B usually moved on foot and by train and wanted to save money, but they also wanted to go to a restaurant with a high rating if possible. For ranks 1, 3, 4, and 5, the extracted connected properties were consistent with the participant's intentions because the participant searched for inexpensive restaurants with high ratings close to a train station. For rank 2, the extracted connected properties showed that the selected properties were low ratings and a wide area. We consider that this connected property was inconsistent with the participant's intentions. Participant B searched for "Japanese food" in task 3. Initially, this task retrieved few results so the search property was relaxed using the rating and distance. This connected property appeared only once. This rule had a high confidence value and a low support value. Therefore, we had to set a threshold for the support value. We consider that this participant sacrificed the rating and distance while retaining the price selection. Therefore, we had to extract connected properties for the price and rating, or the price and distance. We also expected that this operation would include the property priorities. Participant B changed the rating and distance without changing the price. We think that the price property had a high priority in this search context. However, our model could not represent this connected property.

Participant C usually moved on foot, even when the destination was fairly far away, and had an adequate budget for dining. The search intentions of participant C were unclear. Therefore, it was difficult to evaluate the connected properties we extracted. We expect that these rules showed the implicit intentions of participant C.

We confirmed that the extracted connected properties differed according to the operational context and that they represented the search intentions of the users. However, we need to select a threshold for the support value and to determine the priority of each search property. In this experiment, we could not extract the high ranked connected properties with negative correlations. Thus, we need to confirm the effectiveness of connected properties with negative correlations using large experimental datasets.

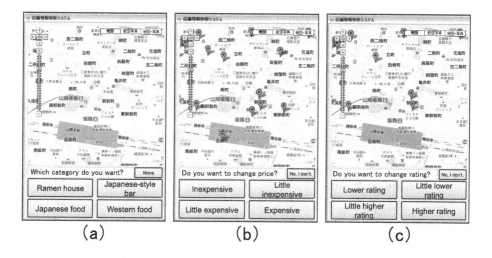

Fig. 4. Example of a restaurant search using our system

5   Application of the Geographical Information Retrieval System

We developed an automatic property adjustment function using the extracted connected properties. This function allows a user to change a property's value based on the automatically extracted connected properties when other property values were changed by the user. The geographical information retrieval system shows the property value options based on the extracted connected properties. When a user selects an option, the system changes the connected properties automatically. This function reduces the number of user operations required to search for geographical information.

This system comprises a search interface, a database of connected properties, a database containing the user's operation history, and a database of restaurant information with geographical information. We describe the interface of our system based on Figure 4. The interface comprises the online map component (the upper part of Figure4) and the operation suggestion component (the lower part of Figure 4). We refer to this as the question mode interface. The users can select only the properties suggested by the system, whereas they can retrieve suitable restaurant information using the online map component. The database of connected properties stores the connected properties extracted from the user operation history in the normal mode. The records maintained in this database comprise the rule ID, user ID, operational context property, operational context value, connected property X, variation of property X, connected property Y, variation of property Y, and association rule score.

Table 4 shows an example of the storage table used by the connected properties database. Rule ID 2 shows a search for "Japanese-style bar." The price and distance are connected properties with a score of 0.33. When the user set a

Table 4. Example of connected properties database

Rule ID	1	2	3	4	5
User ID	A	A	A	A	B
Operational context	Rating	Category	Category	Price	Category
Value of operational context	High	Japanese-style bar	Japanese-style bar	Inexpensive	Ramen house
Connected property X	Price	Price	Rating	Rating	# of displayed objects
Value of connected property X	2000	-1000	-1.0	-1.0	-10
Connected property X	Distance	Distance	Price	Distance	Price
Value of connected property X	-500	300	1000	200	1000
Score of association rule	0.50	0.33	0.24	0.20	0.66

lower price of 1000 yen, the system extended the distance by 300 m. The user's operational history was stored in a database in the normal mode. The restaurant information database stored the restaurant information collected via the Tabelog API.

Figure 4 shows an example of a restaurant search using our system, which is based on the connected properties in Table 4. First, our system extracted four frequent categories from the user's operational history database and these categories are shown as the choices. In this example, the system choices were "Ramen house," "Japanese-style bar," "Japanese food," and "Western food" (see figure 4 (a)). After the user selected a category, the system set the average value of this category for each property based the user's operational history database. Next, the system extracted the connected properties for the user-selected category as the operational context. For example, when user A selected "Japanese-style bar," the system extracted rule ID 2 because this rule contained user ID "A" and the value of the operational context "Japanese-style bar" had the highest scoring association rule. The system also determined the choices when setting the connected property X. In this example, connected property X was the price. Therefore, the system showed the options of "expensive," "moderately expensive," "moderately inexpensive," and "inexpensive" (see Figure 4 (b)). The variation in these choices was based on the variation of property X. In this case, the variation levels were set as "+2000 yen," "+1000 yen," "-1000 yen," and "-2000 yen." After the user selected a choice, the system set the price property and changed the value of connected property Y. In this example, the system automatically adjusted the distance property based on that variation of property Y. The system then extracted the connected properties again with the user-selected choice as the operational context. The system showed other choices based on the extracted connected properties (see Figure 4 (c)). During the repetition of this process, the system adjusted the search properties automatically to reduce the number of user property operations.

6   Conclusion

In this study, we developed a method for extracting the relationships among search properties using an association rule to mine geographical information retrieval systems and to evaluate the extracted connected properties. The connected properties differed according to the user's search context. Therefore, we extracted the connected properties based on operational context with the user's search operation history.

In future work, we aim to improve the extraction method based on association rule mining. We will determine the threshold for the support value and improve the calculation of the score for the association rule by balancing the confidence value and the support value. In addition to association rule mining, we could also Hidden Markov Models and other techniques to extract the connected properties. Thus, we need to compare these techniques. We will also evaluate the effect of reducing the number of user operations with the newly developed system.

Acknowledgment. This research was supported in part by a Grant-in-Aid for Young Scientists (B) 24700098 from the Ministry of Education, Culture, Sports, Science, and Technology in Japan.

References

1. Kobayashi, K., Kitayama, D., Sumiya, K.: Recommender System for Mobile Devices based on Extracting Criteria of Local Search using User's Operation. In: Proc. of the 11th International Working Conference on Advanced Visual Interfaces (AVI 2012), pp. 677–680 (2012)
2. Cho, M., Kim, B., Jeong, D.K., Shin, Y.-G., Seo, J.: Dynamic Query Interface for Spatial Proximity Query with Degree-of-interest Varied by Distance to Query Point. In: Proc. of the SIGCHI Conference on Human Factors in Computing Systems (CHI 2010), pp. 693–702 (2010)
3. Bostandjiev, S., O'Donovan, J., Höllerer, T.: TasteWeights: A Visual Interactive Hybrid Recommender System. In: Proc. of the 6th ACM Conference on Recommender Systems (RecSys 2012), pp. 35–42 (2012)
4. Dalal, O., Sengamedu, S.H., Sanyal, S.: Multi-Objective Ranking of Comments on Web. In: Proc. of the 21st International Conference on World Wide Web (WWW 2012), pp. 419–428 (2012)

5. Liu, L., Mehandjiev, N., Xu, D.-L.: Multi-Criteria Service Recommendation Based on User Criteria Preferences. In: Proc. of the 5th ACM Conference on Recommender Systems (RecSys 2011), pp. 77–84 (2011)
6. Biancalana, C., Flamini, A., Gasparetti, F., Micarelli, A., Millevolte, S., Sansonetti, G.: Enhancing Traditional Local Search Recommendations with Context-Awareness. In: Konstan, J.A., Conejo, R., Marzo, J.L., Oliver, N. (eds.) UMAP 2011. LNCS, vol. 6787, pp. 335–340. Springer, Heidelberg (2011)
7. Baltrunas, L., Ludwig, B., Peer, S., Ricci, F.: Context-Aware Places of Interest Recommendations and Explanations. Journal of Personal and Ubiquitous Computing 16(5), 507–526 (2012)

The Media Feature Analysis of Microblog Topics

Xing Chen[1], Lin Li[2], and Shili Xiong[3]

School of Computer Science and Technology
Wuhan University of Technology
Wuhan, China
{rebecca_lymx,cathylilin}@whut.edu.cn
Communication University of China
slxiong@cuc.edu.cn

Abstract. As microblogging grows in popularity, many research articles are exploring and studying the micro-blogs, especially the English micro-blogging, i.e., twitter. However, Chinese micro-blog service starts rather late and a few research is about its data and characteristics. In this paper, we give out the media feature analysis of microblog topics. Firstly we present our observations tweets and users from Sina by crawling 14 topics and their 74,662 tweets and give out the topic evolution in a certain interval. Then considering the microblogs under a topic exist a lot of redundant information, so in order to reduce the trainning datasets for studying on microblogging , we respectively select different data source as our datasets and give out the evaluation method. We have also studied the microblog semantic extraction based on the topic model of LDA (Latent Dirichlet Allocation). we conclude that the active period of most micro-blog topics is about a month and the out-dated topics will be replaced by the upcoming and related new topics and find that the tweets that appeared in the peak time or the tweets from authenticated users can reflect the whole tweets situation of a topic. However, due to the microblog text is so short that LDA for semantic extraction is not ideal.

Keywords: Microblogging, Media Feature, LDA.

1   Introduction

Microblogging is a new form of communication in which users describe their current status in short posts distributed by instant messages, mobile phones, email or the Web. One of the popular microblogging platforms is Twitter [1]. This model was co-founded by Evan Williams,etc and started in March 2007. The use of Twitter of Obama in his presidential campaign has become an important turning point for the development of the Twitter history. When Michael Jackson died, within one hour,the message has risen up to 65000 on Twitter, which was a peak of development for Twitter. However, Chinese Mirco-blogging service has originated in the English twitter model. The arrival of Sina Micro-blog service marks that micro-blogging has formally crowed into Chinese people's sights. Updates or posts are made by succinctly describing one's current status through a short message that is limited to

B. Hong et al. (Eds.): DASFAA Workshops 2013, LNCS 7827, pp. 193–206, 2013.

140 characters. Topics range from daily life to current events, news stories, and other interests. Instant messaging (IM) tools including QQ and MSN have features that allow users to share their current status with friends on their buddy lists. Microblogging tools facilitate easily sharing status messages either publicly or within a social network.

By May 2012, Sina Micro-blog users has exceeded 300 million. Nowadays micro-blogging as a widely used medium platform, its diverse features meet the people's information, interpersonal information and other aspects of the new requirements. Users usually revolves around a certain hot topic to spread their own opinions. So studying the microblog media feature is becoming more and more important.

To study the the characteristics of Chinese tweets and users and its power as a new medium of information sharing, We have crawled 14 topics and its 74,662 tweets from Sina.We analyze the topic evolution and compare the methods that getting data from different sources, then study the microblog semantic extraction based on the topic model of LDA.

To the best of our knowledge this work is the first study on the Chinese mirco-blog topic sphere. This paper is organized as follows. We first discuss some related works in Section 2. Section 3 describes our topic evolution and compare how to select our studying data source.we study the Selection and Preprocessing of Dataset in Section 4. Last ,we give out Experimental Results and Analysis in Section 5,and conclude our work in Section 6.

2   Related Work

The rising popularity of online social networking services has spurred research into their characteristics and recent work has forayed into characteristics beyond crawled data [2], [3]. Recently, there are a number of articles about exploring and studying the micro-blogging, especially the English microblogging, i.e., twitter. For example, (Krishnamurthy, Gill, & Arlitt, 2008) summarize general features of the Twitter social network such as topological and geographical properties, patterns of growth, and user behaviors. Others such as (Java, et al., 2007), argue from a network perspective that user activities on Twitter can be thought of as information seeking, information sharing, or as a social activity. Newman et al. [4] have made the first quantitative study on the entire Twitter sphere and information diffusion on it. They studied the topological characteristics of Twitter and its power as a new medium of information sharing and have found a non-powerlaw follower distribution, a short effective diameter, and low reciprocity, which all mark a deviation from known characteristics of human social networks. In 2007, Java et al. [5] conduct preliminary analysis of Twitter. They find user clusters based on user intention to topics by clique percolation methods. Krishnamurthy et al. [6] also analyze the user characteristics by the relationships between the number of followers and that of followings. In 2010, Huberman et al. [7] reports that the number of friends is actually smaller than the number of followers or followings. Jansen [8] conducts preliminary analysis of word-of-mouth branding in Twitter.

With the rise of Sina microblogging, more and more researchers works to study the new Chinese micro-blog media. Liu et al. [9] combine a translation-based method with a frequency-based method for keyword extraction. They extract keywords for microblog users from the largest microblogging website in China, Sina Weibo. Our work is also based on the background of chinese Micro-blog topics from Sina. We describe our topic evolution and compare how to select our studying data source , then study the microblog semantic extraction based on the topic model of LDA .

3   Dataset Description

In this section, we first introduce how to extract microblog data by crawling. Then we discuss and analysis the distribution of the tweets and users for each topic in the time interval of 20 days.

3.1   Extracting Microblog Data by Crawling

For benefiting from our professional point of view of computer science technology, we select the technology/IT Internet as our data source. The chosen 14 topics all are the most popular topics at the crawling time as our experimental data. We extract the tweets of these 14 topics. Sina microblog only gives 10 pages space capacity to display the tweets of each micro-topic and each page only exists 20 tweets. From the end of March 2012, we start collecting micro-topic data. In order to ensure nonduplication of data, nearly every two days, we download the html webpages of each Micro-topic, and then save them in the local disk folder as .txt file format. By the end of June 2012, we obtain almost 3750 web pages. But how to extract our needed information from these html files? Here we use HtmlParser[1]. HtmlParser is an open-source project used to parse the HTML document. It is small, fast, simple and has a powerful function.

3.2   Topic Evolution

In order to discuss the distribution of users and tweets in a period of time. In Figure 1, we plot a graph that every 20 days the numbers of users and tweets in a topic.

First, from the overall curve, we can see that at the beginning the number of the tweets and the users is increasing gradually as the time goes on. It means that when a topic is just emerging, it will attract a lot of people. Then, the number of tweets and users reached the maixmum which means the peak period is arrived. After the peak period, the number of users and tweets begin to reduce slowly until the topic die, which means no users are involved in this topic.

---

[1] http://htmlparser.sourceforge.net/

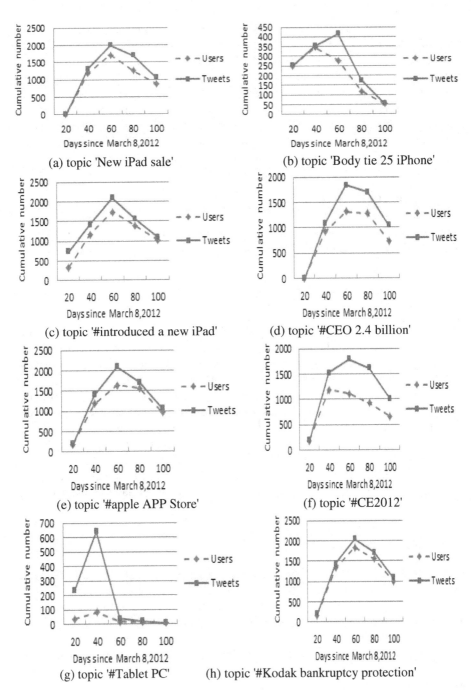

(a) topic 'New iPad sale'

(b) topic 'Body tie 25 iPhone'

(c) topic '#introduced a new iPad'

(d) topic '#CEO 2.4 billion'

(e) topic '#apple APP Store'

(f) topic '#CE2012'

(g) topic '#Tablet PC'

(h) topic '#Kodak bankruptcy protection'

Fig. 1. The Evolution Curves of users and tweets in different topics

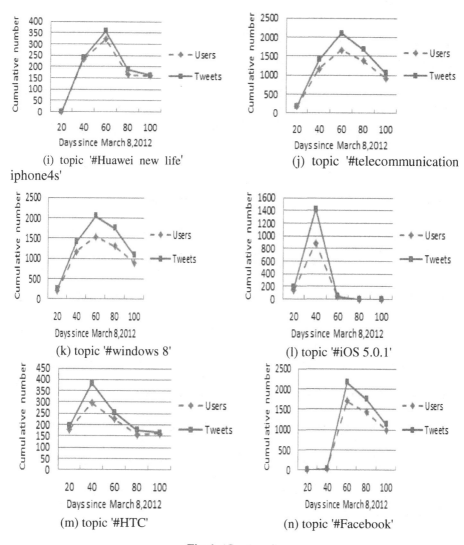

(i) topic '#Huawei new life' iphone4s'

(j) topic '#telecommunication

(k) topic '#windows 8'

(l) topic '#iOS 5.0.1'

(m) topic '#HTC'

(n) topic '#Facebook'

Fig. 1. *(Continued)*

4   Selection and Preprocessing of Dataset

In this section, we first get data source from diffierent aspects.We collected a sample of almost three months tweets between March 25th and June 17th from Sina Microblogging platform, which contains 22,724 tweets from authenticated users and 63,354 tweets from total users, and we also collected tweets that from diffierent period for each of topics. It is worth mentioning that the so-called authenticated users mainly includes the users of Sina microbog VIP, Sina approved Sina agencies and authenticated Sina individuals. Then we give out our approach for text processing and evaluation method. Finally we present semantic extraction for all tweets under 14 topics based on LDA.

4.1   Text Processing and Evalution Method

For each of topics and each of tweets, we conduct the preprocessing of removing stopwords and chinese participle preprocessing. To complete text processing, two NLP tools (i.e.,MyT xtSegT ag andMyZiCiFreq) is used[2]. The next step is to remove all stopwords. We remove the words in our tweets for each topic that appeared in a stopwords list. After making stopwords processing for authenticated tweets, the whole tweets and the tweets that from diffierent period of each topic, we make Chinese participle preprocessing for the filtered tweets by a set of word segmentation and POS tagging tool which is named MyT xtSegT ag. The big advantage of this software is that it can identify proper and newly appeared nouns and minimize the word granularity, such as the new word of Mirco-Letters, weixin(a mobile phone chat software) which is a new application launched by Tencent company in 2011. If the option of starting proper nouns is not selected, then after making participle processing, the word of weixin will be divided into two words that "Micro" and "letters". So using this software can improve the precision and accuracy of participle processing results. Last, we save these produced words in a .txt formatted file, making a preparation for word frequency statistics and analysis. We adopt a word frequency statistic tool named MyZiCiFreq. This software can not only make character frequency statistics but also make word frequency statistics.

Then word frequency statistics are done and top 10 representative nouns or verb-nouns are extracted. In most query suggestion articles, people usually adopt query logs to determine possible query suggestions. But nowadays, with the rise of social network media, many people begin to take microblog as their study background instead of traditional query logs. For a given query, the precision of a query suggestion method is defined as the fraction of suggestions generated that are meaningful. Note that since an exhaustive set of all possible suggestions for a given query is not available, recall cannot be computed. Also, for the query suggestion task, precision is a much more important metric than recall as the number of suggestions that can be offered is limited by the screen space. Here we take our selected top 10 representative nouns or verb-nouns as suggestion list. Hence in order to measure the quality of query suggestion, we decide to select the precision value as our evaluation method which is defined as

$$\text{Precision@N} = \frac{\#\text{related words in a suggestion list}}{N} \tag{1}$$

In Equation 1, we take the extracted top 10 representative nouns or verbnouns as the words that can represent a certain topic. We manually judge whether these words can be considered to accurately reflect the topic. The precision value of each topic is computed.

---

[2] They are recommended by the website of http://www.chinalanguage.gov.cn/index.htm

Table 1. Explanation for the abbreviations in Table 2

	Specific meaning
Authenticated tweets	The tweets that from authenticated users
Total tweets	The tweets that from all of users involved in a topic
Sum1	The sum of tweets that from these period 3.08--3.23,4.10-4.25,4.26-5.11
Authenticated/Total	The tweets that from authenticated users/The tweets that from all of users involved in a topic
Atop10 precision	The precision value of top10 terms that produced by the tweets that from authenticated users
Ttop10 precision	The precision value of top10 terms that produced by the tweets that from all of users involved in a topic
Top10/Top5	The precision value of top10 terms and The precision value of top5 terms

Table 2. The precision values of each topics

Topics	New ipad sale	Iphone news	Ipad show	Apple ceo salary	App Store	CES2012	HTC
Authenticated tweets	2079	471	2420	1831	3005	1866	487
Total tweets	6874	1419	8313	6324	8175	7453	1385
Authenticated/Total	0.3024	0.3319	0.2911	0.2895	0.3676	0.2504	0.3516
ATop10 precision	0.4	0.6	0.4	0.5	0.4	0.4	0.5
TTop10 precision	0.5	0.6	0.4	0.5	0.4	0.3	0.5
ATop5 precision	0.3	0.4	0.3	0.2	0.3	0.3	0.4
TTop5 precision	0.3	0.3	0.3	0.4	0.4	0.2	0.5
Topics	Tablet pc	Kodak bankrupt	Huawei for new life	Iphone 4s sale	Windows 8	iOS jailbreak	Facebook
Authenticated tweets	443	2472	340	2491	2454	604	1761
Total tweets	1556	9084	1448	8085	7125	1932	5639
Authenticated/Total	0.2847	0.2721	0.2348	0.3081	0.3444	0.3126	0.3123
ATop10	0.4	0.4	0.4	0.3	0.5	0.6	0.3

precision							
TTop10 precision	0.4	0.4	0.4	0.3	0.4	0.6	0.3
ATop5 precision	0.4	0.4	0.4	0.3	0.4	0.6	0.3
TTop5 precision	0.4	0.2	0.2	0.3	0.3	0.5	0.2

Topics Top10/Top5	New ipad sale	Iphone news	Ipad show	Apple ceo salary	App Store	CES2012	HTC
3.08--3.23		0.5/0.3	0.3/0.3		0.1/0.1	0.2/0	0.4/0.2
3.24--4.09	0.4/0.3	0.4/0.2	0.1/0.1		0.4/0.3	0.5/0.4	0.6/0.4
4.10--4.25	0.4/0.3	0.6/0.4	0.2/0.2	0.4/0.3	0.4/0.4	0.5/0.3	0.5/0.4
4.26--5.11	0.4/0.3	0.5/0.3	0.1/0.1	0.4/0.4	0.4/0.4	0.5/0.3	0.5/0.4
5.12--5.27	0.3/0.2	0.3/0.3	0.1/0.1	0.5/0.3	0.4/0.4	0.5/0.3	0.4/0.4
5.28--6.13	0.3/0.3	0.3/0.3	0.1/0.1	0.3/0.2	0.4/0.3	0.4/0.3	0.5/0.5
6.14--6.29	0.4/0.3	0.3/0.3	0.1/0.1	0.3/0.3	0.4/0.3	0.4/0.2	0.6/0.4
TTop10 precision/TTop5 precision	0.4/0.3	0.6/0.3	0.4/0.3	0.5/0.4	0.4/0.4	0.3/0.2	0.5/0.5

Topics Top10/Top5	Tablet pc	Kodak bankrupt	Huawei for new life	Iphone 4s sale	Windows 8	iOS jailbreak	Facebook
3.08--3.23	0.4/0.4						0
3.24--4.09	0.4/0.4	0.4/0.3		0.3/0.3	0.4/0.3	0.6/0.5	0
4.10--4.25	0.3/0.3	0.3/0.2	0	0.3/0.3	0.5/0.4	0.6/0.5	0.3/0.2
4.26--5.11	0.4/0.2	0.3/0.2	0	0.3/0.3	0.5/0.4	0.4/0.2	0.3/0.2

5.12--5.27	0.4/0.3	0.3/0.1		0.3/0.3	0.5/0.4		0.3/0.2
5.28--6.13	0.4/0.3	0.3/0.2	0	0.3/0.2	0.5/0.3		0.3/0.2
6.14--6.29	0.3/0.2	0.3/0.1	0	0.3/0.2	0.5/0.3		0.3/0.2
TTop10 precision/TTop5 precision	0.4/0.4	0.4/0.3	0.4/0.2	0.3/0.3	0.4/0.3	0.6/0.5	0.3/0.2
Topics	New ipad sale	Iphone news	Ipad show	Apple ceo salary	App Store	CES2012	HTC
3.24-4.09	482	234	841	0	644	690	239
4.10-4.25	1502	352	1546	1513	1514	1493	303
4.26-5.11	1364	268	1504	1482	1517	1381	183
Sum1	3290	854	3891	2995	3675	3564	725
Total tweets	6874	1419	8313	6324	8175	7453	1385
Sum1/Total tweets	0.4786	0.6018	0.4681	0.4735	0.4495	0.4782	0.5235

Topics	Tablet pc	Kodak bankrupt	Huawei for new life	Iphone 4s sale	Windows 8	iOS jailbreak	Facebook
3.24-4.09	670	645	0	636	780	780	0
4.10-4.25	82	1609	397	1564	1529	852	609
4.26-5.11	17	1424	199	1485	1467	18	1567
Sum1	769	3678	596	3685	3776	1650	2176
Total tweets	1556	9084	1448	8085	7125	1932	5639
Sum1/Total tweets	0.4942	0.4049	0.4116	0.4558	0.5299	0.8540	0.3858

Table 3. The average precision values of all the 14 topics

	Average
Authenticated/Total	0.3038
Sum1/Total tweets	0.4665
Atop10 precision	0.4357
Ttop10 precision	0.4286
Atop5 precision	0.3286
Ttop5 precision	0.3214

4.2   Semantic Extraction Based on LDA

LDA is a hierarchical probabilistic model of documents. Here we utilize a C implementation of variational EM for latent Di richlet allocation (LDA), a topic model for text or other discrete data. It allows us to analyze of corpus, and extract the topics that combined to form its documents. The following is steps we take when doing experiment with LDA-c code:

1. Transform original data into required data format.

During the process of analyzing our dataset(14 topics,74,662 tweets), the term index and term-document matrix would be created, which provide great convenience for the transformation of all the original tweets of 14 topics into the data format that LDA-c implementation required. The final data is a file where each line is of the form:
[M] [term_1]:[count] [term_2]:[count] ...   [term_N]:[count]
where [M] is the number of unique terms in one of 14 topics, and the [count] associated with each term is how many times that term appeared in this topics. Note that [term_1] is an integer which indexes the term; it is not a string.

2. Topic estimation and inference

After compiling the code, we could estimate the model by executing:
lda est [alpha] [k] [settings] [data] [random/seeded/*] [directory].
To perform inference on a different set of data execute:
lda inf [settings] [model] [data] [name]
Variational inference is performed on the data using the model produced by estimation. After inference, [name].gamma will be created, it's the variational Dirichlet parameters for each topic, with which the top N words from each topic in a .beta file being printed.

3. Printing topics
Here we can use the Python script topics.py to print out the top N words from each topic in a .beta file. The usage is as following:

python topics.py <beta file> <vocab file> <n words>

In our experiment, usually we take the value of alpha as 1. Parameter k indicates that program according to these 14 topics extract and statistic the k hottest topics. Then through taking the diffierent k value,we can see the distribution of 14 topics under the k hottest topics. Finally we select the topic distribution that probability greater than 1,and print out top 5 and top 10 words from each topic for 14 topics. The experimental results were shown in Table4.

Table 4. ,when k=5, probability>1,top 5 and top 10 words   from each topic for 14 topics

K=5	probability> 1	Top5	Top10
CE20 12	0 1 2 3 4	Phone, Product, America, Telecom, China	Phone, Product, America, HK, Kodak, Telecom, Phone, Company,Software, Release, China
Faceb ook	0 1 2 3 4	Phone, Product, America, Telecom, China	Phone, Product, America, HK, Kodak, Telecom, Phone, Company, Software, Release,China
IOS jailbre ak	0 1 2 4	Phone, Product, America, Telecom, China, Phone	Phone, Product, America, HK, Kodak, Telecom, Company, Software, China, Release
Huaw ei for new life	1 2 3 4	Telecom, China, Phone	Telecom, Phone, Company, Software, Kodak, Release, China
HTC	0 1 2 3 4	Phone, Product, America, Telecom, China	Phone, Product, America, HK, Kodak, Telecom, Phone, Company,Software, Release, China
Windo ws 8	0 2 3 4	Phone, Product, America, Telecom, China	Phone, Product, America, HK, Kodak, Telecom, Software, China, Release
Kodak	0 1 2 3 4	Phone, Product, America, Telecom, China	Phone, Product, America, HK, Kodak, Telecom, Phone, Company,Software, Release, China
Teleco m	0 1 2 4	Phone, Product, America, Telecom, China	Phone, Product, America, HK, Kodak, Telecom, Company, Software, China, Release

Table 4. *(Continued)*

Apple Store	0 1 2 3 4	Phone, Product, America, Telecom, China	Phone, Product, America, HK, Kodak, Telecom, Phone, Company,Software, Release, China
New IPAD	0 1 2 3 4	Phone, Product, America, Telecom, China	Phone, Product, America, HK, Kodak, Telecom, Phone, Company,Software, Release, China
New IPAD Sale	0 1 2 3 4	Phone, Product, America, Telecom, China	Phone, Product, America, HK, Kodak, Telecom, Phone, Company,Software, Release, China
Apple ceo salary	0 1 3 4	Phone, Product, America, Telecom, China	Phone, Product, America, HK, Kodak, Telecom, Phone, Company, Release,China
Iphone news	0 1 3 4	Phone, Product, America, Telecom, China	Phone, Product, America, HK, Kodak, Telecom, Phone, Company, Release,China
Talet PC	0 3 4	Phone, Product, America, Telecom, China	Phone, Product, America, HK, Kodak, Telecom, Phone, Release, China

5   Experimental Results and Analysis

Here, in Table 1, we have made a specific explanation for all abbreviations appeared in Table 2 which show the results of each microblog topic. The average results of all the 14 topics are listed in Table 3. From the result that presented in Table 2, we can clearly see that the differences between Atop10 precision and Ttop10 precision are not particularly obvious in terms of precision. For further observation, we decide to make an average for the precision values of 14 topics. To our surprise, the precision value of top 10 words that produced by the tweets that from authenticated users is higher than that produced by the tweets from all of users involved in a topic on average. At the beginning, we think that the number of the total tweets for one topic actually not only contains the tweets from authenticated users, but also contain others tweets from common users. In comparison, it has the larger suggestion context source and the richer content information. Thus, it should output higher precision scores. However, the results run adversely to what we might intuitively expect the average precision value of top 10 words that produced by the tweets that authenticated users, i.e., slightly higher.

However, we also do the job of computing the top10 and top5 percision value of the tweets that from different periods. By comparing with the value of both TTop10 precision and TTop5 precision, we find that the highest persion value are most concentrated in the 3.24 to 5.11,according to the curves of topic evolution ,they are appeared around the peak time.

So what does this show? It illustrates that we need not to select all tweets of a topic as our suggestion context source. Considering the final result, the tweets that from authenticated users and the tweets that around the peak time could be on behalf of the entire tweets under a topic. During the preprocessing, we observed that under the background of computer configuration with a 32-bit operating system, dual-core CPU and 3.00GB memory, it takes about 4 hours for all users' tweets of a certain topic, but for processing authenticated users' tweets, it just takes about 30 minutes. Taking tweets that from authenticated users as our suggestion context source saves not only the processing time, but also the storage space. How much storage space does it save at all? From experimental data, we can see that average Authenticated/Total is about 0.3038 and Sum1/Total tweets is 0.4665. In other words, the authenticated context accounts for around 1/3 in total tweets and almost save 2/3 storage space. The tweets that appeared around the peak time accounts for around 1/2 in total tweets and almost save 1/2 storage space.

In Table 4,we can see that the top5 and top 10 words do not reflect some topic. Due to the microblog text is too short, making the analysis of the hiden topic analysis for the tweets does not seem ideal.

6   Conclusion

In this paper, firstly we introduced how to extract microblog data by crawling. Then we discussed and analysised the distribution of the tweets and users for each topic in the time interval of 20 days. Then we gave out our selection of datasets and the approach for text processing and evaluation method. Finally we presented semantic extraction for all tweets under 14 topics based on LDA. Considering the final results, we can see that the tweets that from authenticated users and the tweets that around the peak time could be on behalf of the entire tweets under a topic. Due to the microblog text is too short, making the analysis of the hiden topic analysis for the tweets does not seem ideal. In the further, We will combine the media characteristic of mircoblog to da some jobs of social network, such as query suggestion and so on.

References

[1] Pontin, J.: From many tweets, one loud voice on the internet. The New York Times (April 22, 2007)
[2] Benevenut, F., Rodrigues, T., Cha, M., Almeida, V.: Characterizing user behavior in online social networks. In: Proc. of ACM SIGCOMM Internet Measurement Conference. ACM (2009)

[3] Wilson, C., Boe, B., Sala, A., Puttaswamy, K.P., Zhao, B.Y.: User interactions in social networks and their implications. In: Proc. of the 4th ACM European Conference on Computer Systems. ACM (2009)

[4] Newman, M.E.J., Park, J.: Why social networks are different from other types of networks. Phys. Rev. E 68(3), 036122 (2003)

[5] Java, A., Song, X., Finin, T., Tseng, B.: Why we twitter: understanding microblogging usage and communities. In: Proc. of the 9th WebKDD and 1st SNA-KDD 2007 Workshop on Web Mining and Social Network Analysis. ACM (2007)

[6] Krishnamurthy, B., Gill, P., Arlitt, M.: A few chirps about twitter. In: Proc. of the 1st Workshop on Online Social Networks. ACM (2008)

[7] Huberman, B.A., Romero, D.M., Wu, F.: Social networks that matter: Twitter under the microscope. First Monday 14(1) (2009)

[8] Jansen, B.J., Zhang, M., Sobel, K., Chowdury, A.: Microblogging as online word of mouth branding. In: Proc. of the 27th International Conference Extended Abstracts on Human Factors in Computing Systems, pp. 3859–3864 (2009)

Web Service Based Algorithm Management Framework for Stream Data Processing

Xiaodong Zhu and Hengshan Wang

Department of Information Management,
University of Shanghai for Science and Technology,
Shanghai, China, 200093
zhuxd@usst.edu.cn, wanghs@usst.edu.cn

Abstract. In recent ten years, In contrast to lots of work on developing algorithms for stream data mining, little work has been done in the management of various stream data processing algorithms, which make the application rate of algorithms quite low. In this research, we first establish a process model for data stream mining. An algorithm management framework based on web services for stream data processing, AMF4SDP, is then proposed. We analyze the construction of data stream processing algorithm repository and the architecture of the framework. Using the framework, a data stream oriented algorithm management prototype system is implemented on Eclipse. Experiments validate that the framework has high flexibility and self adaptability.

Keywords: Stream data processing, Process Model, Data Streams, Predictive Model Markup Language, Web Service.

1 Introduction

Data streams are a kind of project sequences which appears steadily and ceaselessly. Meanwhile, data streams are continuous, substantive as well as limitless compared to the traditional static data. Moreover, data streams usually appear with super speed and the characteristics of whose data distribution is changing with time variation. The application of data streams in many fields make the study of data stream mining a hot focus, which includes the application domains of financial stock market, network traffic surveillance and control system, sensor network, the telephone log of the telecom sector and so on.

In recent years, a series of algorithms of data stream mining have been put forward [1][2][3][4][5][6][7], which are effective when they solve the specific issues in the given application fields. As a whole, the data stream mining algorithms can be divided into regression analysis, classification analysis, cluster analysis and association analysis, etc. Nevertheless, for each data stream mining algorithm, it solves a data stream mining issue only lies in a given condition or in a certain area. It requires different algorithms for various data stream or dissimilar periods of data mining. Facing the new requirement, we need to develop a totally new algorithm. Meanwhile, repeated experimentation must be carried out, which

B. Hong et al. (Eds.): DASFAA Workshops 2013, LNCS 7827, pp. 207–219, 2013.

leads to the low algorithm availability. It has been the new requirements for the development of the data stream mining technique to understand and solve the issue of combining multitudinous data stream mining algorithms and structuring rational model library of data stream mining, which ultimately reducing the repeated development work of data stream mining algorithms.

Compared with much work on developing algorithms of data mining, there is little work paid on the nature, theoretic foundation, services, platforms and stands of data mining[8][9][10][11]. With the development of the data mining technique, the standardization of data mining has become the cause of anxiety day by day. In the recent several years, the KDD international conference hold by ACM has entered into the standardization of data mining as the theme and symposium. Many industrial standard for data mining has been proposed, such as the standard procedure CRISP-DM of multi-industry data mining[12][13], the markup language PMML of prediction model and so on. CRISP2DM provides a kind of process standard that is a description of the whole life cycle of data mining, which has currently been a standard method used in the process of developing data mining projects. PMML is a sort of modeling language of the standard process model for data mining, which enables to share the models among each system. The Microsoft Corporation incorporates the PMML standard which results in the specification language of data mining called OLE DB for data mining, including creating primitive and many other significant definition and usage of data mining models.

Currently, some researches on traditional process model of data mining have already been carried on. Romei et al put forward the whole process descriptive language KDDML based on the XML mining model[15], which makes it convenient for algorithms added as modules to the process model. Yang adopted the meta-learning method to choose the suitable mining algorithm and come up with the quality essence of mining system[16]. Cheung and the rest buried themselves in their study of framework for distributed data mining of service architecture[17]. Perttu and the rest studied the framework for data mining based on component technology, structuring the mining model applied by the mining algorithms as the components[18]. However, the framework of this kind does not satisfy the proper adaptability. Liu proposed an implementation approach for data mining model PMML based on work flow[19].

2   Process Model of Data Streams

CRISP2DM organizes a process model used to represent data mining via developing the life circle of data mining project. Based on the characteristics of data stream mining, the Figure 1 below describes the process model for data stream mining. The square frame represents each phase; the arrows represent the direction of flow and the dependency relationship between two phases. Data stream is the center of the whole mining process. Understanding a problem is the initial phase. The last phase is the maintenance of the module. If there exists a relationship of repeat or feedback between two processes, the symbol of a double-headed arrow is represented. The process model of data stream mining PM4DSM generalized by an eight tuple showed as follows.

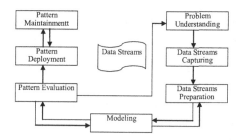

Fig. 1. Process Model for Data Streams Mining

Definition 1. $PM4DSM =< DS, PU, DC, DP, M, PE, PD, PM >$

And DS means data stream, PU, DC, DP, M, PE, PD, PM are the seven phases in file, namely problem understanding, data streams capturing, data streams preparation, modeling, pattern evaluation, pattern deployment and pattern maintenance. The paper analyzes the process model of data stream mining from the phase PU to the phase PM.

The data of the data streams is called the stream data. In fact, data stream DS can be defined as a set of two-tuples data.

Definition 2. $DS = \{< d, \tau > | d \models D, \tau \models T\}$

where d is the record that satisfies data set format D. τ is the time stamp of data recording, which satisfies the time format T. It indicates that data streams are ordered item sequences whose time series is emerging constantly.

Compared to the traditional static data analysis, the data stream mining has following salient characteristics: first of all, the input data processed by data stream is abundant and sequential data stream which is in random occurrence instead of being fixed on disks or memorizers. Secondly, the size of the data stream is potentially infinite. Therefore, it is a mistake to store all the stream data in main memory or external memory. Furthermore, it is emerging continuously for data stream as well as data distribution characteristics. Therefore, it is crucial to have a real-time update aiming at the results of data stream mining, to provide the sequential consequence. Finally, due to the faster speed at which data stream arrives, consequently it requires higher requirements to data stream mining algorithms. For example, it claims to get the result after one data scanning rather than repeated scanning procedures. In addition, data stream mining takes more system resource consumption into account, such as CPU, memory space, so far as to energy restriction and the other things.

3   Algorithm Management Frameworks for Data Stream Mining

3.1   Framework Architecture

There exists many corresponding algorithms in data streams capturing, data preparation and modeling phase, especially in the data stream mining phase. Compared to traditional static data, the context factor of data streams makes a claim for higher requirements to data stream mining algorithms. It requires various data stream mining algorithms for different data stream domain and data streams preparation pattern. There are many data stream mining algorithms at the present time, but the availability of algorithms is low, which only solve the problem under the specific condition or specific domain, forming train of thought of structuring data stream mining algorithm library. With the development of the internet, XML technique, web service has gradually become mature network techniques. The correlative standard of web service is being put forward and updating continuously. We propose the process model algorithm management framework for streams data processing based on web service technique (AMF4SDP), Showed in Figure 2 below.

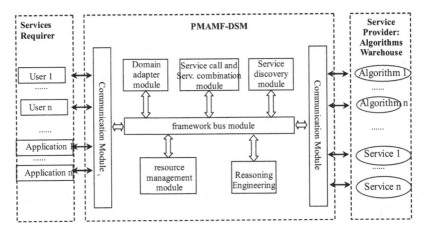

Fig. 2. Algorithm Management Framework for Data Streams Mining

The framework bus module is the pivot of AMF4SDP in Figure 2, used to manage other modules of AMF4SDP. It is mainly responsible for calling other modules and communicating with the modules. Other modules can be dynamically mounted to the bus module, or uninstall from the bus module. The resource management module is used to realize the information storage of AMF4SDP, including the semantic description of information established for web service, events and message information as well as the WSDL and SOAP information which is used to carry through web service call, which are all in line with AMF4SDP.

Service discovery module finds out the web service which fulfils the needs according to the task provided by service requester. Domain adapter module is responsible for the mapping in specific domains and abstract web service, which enables the web service to be reused in certain domain, in order to solve specific problems in domain. Service call module is to realize the interaction between service requester and web service according to the description of choreography in AMF4SDP. Service combination module is to realize the combination which is carried through for completing certain task during the service, according to the description of orchestration in AMF4SDP. All the reasoning tasks in AMF4SDP are completed by inference engine. AMF4SDP has several different modeling languages to choose as a kind of description. Different modeling languages are corresponding to different inference engine. For example, we can choose the corresponding inference engine Jess if we choose OWL as the language to describe the semantic web service. Communication module is responsible for providing the communication between AMF4SDP and algorithm service requester and algorithm service provider, mainly realizing the construction and analysis of message in the communication process. The table 1 defines the interface semantics.

Table 1. Interface Semantics for AMF4SDP

Interface function	Interface semantics
Service getting	Context *GetWebService*(DataStreamsMining *task*)
Service invoking	Context *InvokeWebService*(Context *context*)
Service discovering	WebService *Discover*(DataStreamsMining *Request*)
Domain mapping	Mapping *DomainAdapt*(WebService *webService*, Domain *domain*)
Registering Choreography	void *RegisterChoreography*(WebService *webService*)
Registering Orchestration	void *RegisterOrchestration*(WebService *webService*)
Result getting	Context *GetResult*(WebService *webService*, Context *context*)

In AMF4SDP, data stream mining algorithm library is worked as the service provider, and at the same time, the data stream mining algorithms are released as web service. In order to realize the automatic discoverycall and combination of data stream mining algorithm, the standard for the interface of data stream mining algorithm is a must. Under the PMML standard, the top-level element design of the PMML configuration file in data stream mining algorithm library is shown in figure 3.

Perceived from the figure 3, more than one data stream mining algorithm pattern can be involved in the PMML file. And the pattern is presented by exclusive name. Each algorithm pattern provide detailed OWL language description of input, output and algorithm execution interface, in order to facilitate AMF4SDP carrying out discovery, matching, calls and combination of algorithms.

```
<xs:element name="PMML">
  <xs:complexType>
      <xs:sequence>
        <xs:element ref="DataDictionary"/>
        <xs:choice>
          <xs:element ref="DataStreamsFrequentSetModel"/>
          <xs:element ref="DataStreamsClusteringModel"/>
          <xs:element ref="DataStreamsRegressionModel"/>
          ......
        </xs:choice>
        ......
      </xs:sequence>
  </xs:complexType>
</xs:element>
```

Fig. 3. Predictive Model Markup Language

3.2   The Characteristics of Framework

AMF4SDP possess the characteristic of high adaptability. Service requesters can be users as well as the application programs those need to call data stream mining algorithm. Service requesters establish the data steam mining task at first, establishing the task of data stream correlation analysis, for example. Service requesters pass context parameter to AMF4SDP, such as the data stream rate, size of sliding window, function of data stream mining pattern and so forth. The AMF4SDP domain adapter module may acquire corresponding specific domain by analyzing context parameter.

There exist a lot of algorithms in the process of data stream mining, including the flow rate of data stream, size of sliding windows, which have a significant impact on algorithm performance. AMF4SDP framework collects multiple algorithms used in the process of data stream mining as service and builds the data stream mining algorithms library, which is applied by the web service technique. When users or application programs make corresponding algorithm request, the most appropriate algorithm may be returned to users via AMF4SDP. User receive algorithm can perform the corresponding mining task locally as well. As a result, AMF4SDP is able to improve the utilization of the data stream mining algorithms and choose the most appropriate algorithm according to the environment context of data stream mining, which finally improves the efficiency of the mining task.

4   Experimental Analysis and Evaluation

The section 2 introduces the process model algorithm management framework for streams data processing (AMF4SDP), module functionality and interface semantics. This section will describe implementations of the framework by way of specific experiments. And then it illustrates that the application of AMF4SDP can realize the optimized choice of the process model algorithm of streams data processing by way of the comparative experiments conducted by multiple join query algorithms. Experimental platform adopts Eclipse with Java 5.0 development. The experimental condition configuration is Windows XP operating system with CPU Pentium 1.7G, internal storage 256M and a database using Microsoft SQL Server 2000.

4.1 AMF4SDP Execution Semantics

AMF4SDP framework is used to help service requester find and call related streams data processing algorithms. Figure 4 shows UML activity diagram provides the internal execution semantics based on algorithm management system of AMF4SDP framework. We can see from the figure 4 that AMF4SDP receives the task established by service requester at first. Then it searches algorithms from algorithm library by calling the service discovery module in line with the task. An error will be returned if a suitable algorithm has not been found, otherwise the domain adapter module will be called. Domain adapter module maps the discovered algorithms to the domain of current task. Two cases of success and failure are also involved in the adaptation results. If it succeeds, the call and combination is carried through by way of calling service and combining modules. At the end, the service requester will be returned the combination algorithms.

The experiment deploys algorithm service system on the server and develops the client as a service requester to send request algorithm tasks.

4.2 Algorithm Optimization and Selection

Data stream query is usually an indispensible means of modeling of data stream mining. The data flow rate and the size of sliding windows have greater impact on data stream query performance. So take the data stream join query as an example here, studying the optimization and selection problem of data stream join query algorithms. The data join query devise includes two and a half semi-join algorithms and 3 full-join algorithms. There are two balancing algorithms and one unbalanced algorithm in all three full-join algorithms. It is specifically the HJ semi-join algorithm, NLJ semi-join algorithm, HJ full-join algorithm, NLJ full-join algorithm and HNJ full-join algorithm. Here HJ means using Hash table data structure; NLJ means using linked list data structure; HNJ only aims at full-join ones, indicating one data stream is using Hash data structure while the other one is using linked list data structure. The so-called semi-join just means caring about the result of the data in data stream A connects with the data in data stream B, rather than the result of the connection of the data in data stream B with the data in data stream A. The concept of the full-join indicates the necessity of inquiring the result of the connection between data stream A and data stream B.

Kang is studying the method that how to select join query algorithm in line with the cost model[14]. We try to realize these five join algorithms on the

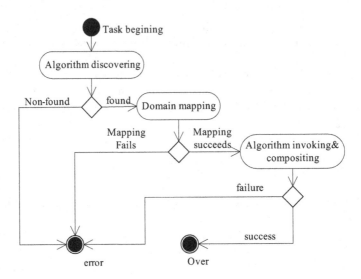

Fig. 4. Activity Diagram in UML

basis of public data sets, deploying these algorithms on the server of AMF4SDP algorithm service system and analyzing merits and demerits of various algorithms in line with the size of windows and data flow rate. Algorithm selection strategy will be deployed on domain adapter modulewhich is used for selecting the optimal join algorithm in specific environment. The experimental data come from the data in table customer on TPC2H test platform (*http://www.tpc.org/tpch*). The two data streams separately use two processes responsible for reading data tuples from table to sliding windows continuously. The flow rate of data stream A is 1000 tuples/s. The flow rate of data stream B is controlled according with A's ration of flow rate. Take NLJ semi-join and HJ semi-join as an example, setting the size of sliding windows 100, number of Hash bucket 10, the total data number of two data stream 1000. The consuming time of two algorithms is as shown in table 2, according to the arrival rate of different data streams.

The conclusion can be drawn from table 2: the ration of the transmission rate of data stream has impact on the execution performance of different semi-join algorithms. And meanwhile the time consumption of semi-join algorithm may increase as the increase in the speed gap of data stream. Due to limited space, the full-join algorithm performance comparison is omitted here. Here lists the selection strategy of join query algorithm according to the algorithm performance, as shown in table 3.

Besides the comparison of semi-join algorithm performance, the table 3 also includes the comparison of full-join algorithm performance. We deploy the algorithm selection strategy formed in table 3 on domain adapter module of service system.

When the experiment sends the task of the request service of data stream semi-join algorithm from the server to AMF4SDP algorithm service system,

Table 2. Half Joins Algorithms Performance of Data Streams

algorithm / speed propotation	NLJ（scale：millisecond）	HJ（scale：millisecond）
1:1	68565	67967
1:2	71782	69383
1:3	80426	70328
1:4	81156	72258
1:5	84368	76213
1:6	89025	80476
1:7	89822	87527
1:8	90337	91915
1:9	91533	98038
1:10	92762	141769

Table 3. Half Joins Algorithms Performance of Data Streams

Speed Proportion of Stream data	Join algorithms performance
1:1~1:7	HJ semi-join is better than NLJ semi-join
>=1:8	NLJ semi-join is better than HJ semi-join
1:1~1:2	HJ full-join is best among full-join algorithms
>=1:3	HNJ full-join is best among full-join algorithms

it sends the flow rate of data stream "1:4"and "1:20"and the task parameter "Data Streams Query" one after another. And the result is that the experiment acquires respectively the modules and interfaces of HJ semi-join algorithm and NLJ semi-join algorithm returned from AMF4SDP algorithm service system. The client continues to carry through the join query operation after acquiring these two algorithms.

The flow rate of data streams and the size of sliding windows are the context parameters of data streams in AMF4SDP algorithm service system. When sending the join query commands of data streams, these commands will be sent to the algorithm service system together with the context parameters. Then the context parameters will be analyzed by the domain adapter module. In addition, the service discovery module will find the most appropriate join query algorithm of data stream. The concerned algorithms will be combined by service calling combination modules. Finally, the algorithms will be returned to the client to carry out join and query. The experiment proceeds to study the comparison of the service efficiency of using AMF4SDP framework with the efficiency of not using it. It is assumed that the client has possessed the semi-join query algorithm of data stream, or the client complexity of executing the task. When the AMF4SDP framework is no more used, the initial algorithm will be always used because the client cannot determine to select which join query algorithms to use. In experiment, the HJ semi-join algorithm is used. When the flow speed ratio of data streams is above 1:8, the efficiency of query task will also be reduced correspondingly based on the low efficiency of the semi-join query algorithm. Obliged to the merits of algorithm management system of AMF4SDP framework, that it in advance deploys concerned algorithms of the process model of data stream mining on the system by way of web service, provides the corresponding interface; system returns the optimization algorithm matching the context parameter of users when users and application programs send request algorithms, which result in the convenience and efficiency for users to acquire the algorithms and execute the corresponding mining task immediately in the client.

We can deploy the algorithms selecting strategy into domain adapter module of AMF4SDP. The composition algorithm service process is created as follows figure. 5.

The composition service is named as "Stream data query algorithm service". Arrow lines in the figure represents the service process. The "Start/In" denotes the beginning of the process, "Finish/Out" denotes the finishing of the process. "P" indicates that the process has results. The process has input data and output data . "Get Stream Data Context" is an atomic service, whose input and output are designed by users. "Judge Stream Speed?" is the condition judgement expression, which is used for judging the value of the atomic service "Get Stream Data Context". The partial semantics of the figure is represented in PMML language as follows.

```
<algorithm_process:CompositeProcess rdf:ID="FrequentItemsetsMining_Process">
  <algorithm_service:describes rdf:resource="#FrequentItemsetsMining_Service"/>
  <algorithm_process:hasInput rdf:resource="#Support_Threshold"/>
  <algorithm_process:hasOutput rdf:resource="#FrequentItemsets_Results"/>
  <algorithm_process:composedOf>
  <algorithm_process:Sequence rdf:ID="Sequence_1">
```

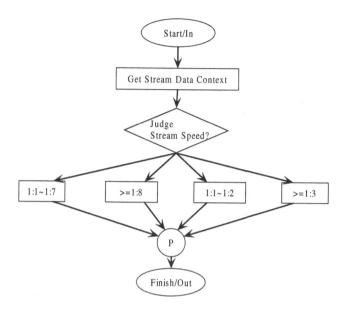

Fig. 5. Composition Algorithm Service Process

```
<algorithm_process:components>
 <algorithm_process:ControlConstructList rdf:ID="ControlConstructList_1">
  <algorithm_list:first>
   <algorithm_process:Perform rdf:ID="Perform_StreamContext">
    <algorithm_process:process>
     <algorithm_process:AtomicProcess rdf:ID="GetStreamContext_Process">
      ...
     </algorithm_process:AtomicProcess>
    </algorithm_process:process>
   </algorithm_process:Perform>
  </algorithm_list:first>
  <algorithm_list:rest>
   <algorithm_process:ControlConstructList rdf:ID="ControlConstructList_2">
    <algorithm_list:first>
     <algorithm_process:If-Then-Else rdf:ID="If-Then-Else_1">
      ...
     </algorithm_process:If-Then-Else>
    </algorithm_list:first>
    <algorithm_list:rest>
     ...
    </algorithm_list:rest>
   </algorithm_process:ControlConstructList>
  </algorithm_list:rest>
 </algorithm_process:ControlConstructList>
 </algorithm_process:components>
 </algorithm_process:Sequence>
 </algorithm_process:composedOf>
</algorithm_process:CompositeProcess>
```

5   Conclusion

Data stream mining has been a hot research in data mining domain due to a wide
range of its application. However, the low utilization factor and fewer studies for

the process model of data stream mining make it a problem in spite of the current multiple algorithms of data stream mining. The main tasks of this paper are: (1). Expanding the process model of data mining to data stream mining; putting forward and analyze the process model of data stream mining in detail; discussing the seven phases of data stream mining. (2). Putting forward the process model algorithm management framework for streams data processing based on web service technique (AMF4SDP); discussing the structure of AMF4SDP algorithm library; analyzing the adaptability of AMF4SDP. The mentioned framework in this paper helps understand the process of data stream mining, improve the availability of data stream mining algorithms and reduce the duplication of labor and experiments of developing data stream mining algorithm.

This paper has achieved a simple algorithm management system- AMF4SDP under the Eclipse developing environment. The result of experiments indicates that the AMF4SDP framework possesses the high adaptability in discovery, call and combination of data stream mining algorithms. The further work includes establishing and realizing a more perfect process model algorithm management system of data stream mining. It is necessary to further study the problem of the algorithm search strategy in data stream mining algorithm library, the comparison of the similarity of algorithms and the optimal combination of algorithms.

Acknowledgement. This research is supported by the Humanity and Social Science Youth foundation of Ministry of Education of China under Grant No.12YJC870037, the National Natural Science Foundation of China under Grant No.71071098, the Research Fund for the Doctoral Program of Higher Education of Ministry of Education under Grant No. 20123120120004, the Innovation Program of Shanghai Municipal Education Commission under Grant No. 12YZ103, and the Excellent Youth Scholars of Ministry of Education of Shanghai under Grant No. slg10010.

References

1. He, H., et al.: Incremental learning from stream data. IEEE Transactions on Neural Networks 22(12, pt. 1), 1901–1914 (2011)
2. Lian, X., Chen, L.: Similarity join processing on uncertain data streams. IEEE Transactions on Knowledge and Data Engineering 23(11), 1718–1734 (2011)
3. Guha, S., Meyerson, A., Mishra, N., Motwani, R., O'Callaghan, L.: Clustering Data Streams: Theory and Practice. TKDE Special Issue on Clustering 15 (2003)
4. Indyk, P., Koudas, N., Muthukrishnan, S.: Identifying Representative Trends in Massive Time Series Data Sets Using Sketches. In: The 26th International Conference on Very Large Data Bases, Cairo, Egypt (2000)
5. Manku, G.S., Motwani, R.: Approiximate frequency counts over data streams. In: 28th International Conference on Very Large Data Bases, Hong Kong, China (2002)
6. Wang, H., Fan, W., Yu, P.S., Han, J.: Mining Concept-drifting Data Streams using Ensemble Classifiers. In: The 9th ACM International Conference on Knowledge Discovery and Data Mining (SIGKDD), Washington DC, USA (2003)

7. Zhu, X., Xiao, F., Huang, Z., Shen, G., Jin, L.: Description Logic Based Extended Predictive Model Markup Language EPMML. Chinese Journals of Computers 35(8), 1644–1655 (2012) (in Chinese Language)
8. Zhu, X., Wang, H., Gan, H., Gao, C.: Construction and Management of Automatical Reasoning Supported Data Mining Metadata. In: 2011 IEEE International Conference on Supernetworks and System Management, pp. 205–210 (2011)
9. Zhu, X., Yang, J.: An Extended Predictive Model Markup Language for Data Mining. In: Chen, L., Tang, C., Yang, J., Gao, Y. (eds.) WAIM 2010. LNCS, vol. 6184, pp. 218–231. Springer, Heidelberg (2010)
10. Zhu, X.-D., Huang, Z.-Q.: Conceptual modeling rules extracting for data streams. Knowledge-Based Systems 21, 934–940 (2008)
11. Zhu, X., Huang, Z., Shen, G.: Description Logic based Consistency Checking upon Data Mining Metadata. In: Wang, G., Li, T., Grzymala-Busse, J.W., Miao, D., Skowron, A., Yao, Y. (eds.) RSKT 2008. LNCS (LNAI), vol. 5009, pp. 475–482. Springer, Heidelberg (2008)
12. CRISP-DM.:CRoss Industry Standard Process for Data Mining (2011), http://www.crisp-dm.org
13. DMG.:PMML Version 3.1. (2010), http://www.dmg.org/index.html
14. Kang, J., Naughton, J.F., Viglas, S.D.: Evaluating Window Joins over Unbounded Streams. In: Proceedings of the 28th VLDB Conference, Hong Kong, China (2002)
15. Romei, A., Ruggieri, S., Turini, F.: KDDML:A middleware language and system for knowledge discovery in databases. Data & Knowledge Engineering 57, 179–220 (2006)
16. Yang, L., Zuo, C., Wang, Y.G.: Research and Implementation of Service Oriented Architecture for Knowledge Discovery. Chinese Journal of Computers 28, 445–457 (2005)
17. Cheung, W.K., Zhang, X., Wong, H., Liu, J.: Service-Oriented Distributed Data Mining. IEEE Internet Computing (2006)
18. Lauinen, P., Tuovinen, L., Ring, J.: Smart Archive: A Component-based Data Mining Application Framework. In: The 5th International Conference on Intelligent Systems Design and Applications (2005)
19. Liu, G., Yuan, S., Dong, L., Li, Y.: An Implementation of Data Mining PMML Based on Work Flow. Journal of Chinese Computer Systems 28, 891–894 (2007)

Research on Petri Net and IDEF1X Based Production Scheduling Modeling Method

Qin Jiang-tao and Ren Shang

University of Shanghai for Science and Technology, Business School, Shanghai, 200093
qinjiangtao_usst@126.com

Abstract. A systematically design color sets of colored Petri net by applying IDEF1X method is proposed in this paper. Through analyzing the current deficiency of introducing time factor into Petri net, the time extension mode of timestamp in Petri net is given in this paper. By using the method mentioned above, a production scheduling model with great generality is established, and the modeling system performs well in practical applications.

Keywords: System Modeling, Timed Colored Petri Net, IDEF1X, Production Scheduling Model.

1   Introduction

Petri net is used to describe a kind of model of a distributed system, it is able to describe the structure of a system as well as simulate the operation of the system, and it offers richer model information than other modeling tools. Nowadays, Petri net is more widely applied in manufacturing systems on scheduling, control modeling and performance analysis. But when it was used in large-scale complex systems, there are still lack of the following: 1, the existing node is too much and the system is too complex when the system is build on the basic Petri nets, it will be difficult and time consuming to understand; 2, early Petri net theory have no idea with time concept, it can't describe temporal activity and difficult to show when an event happened and the consumption of time.3, it's ability is limited, such as lack of modular programming, hierarchical modeling and so on [1, 2].

Therefore, people expand the basic Petri nets in order to strengthen the Petri net modeling and analysis ability, puts forward a variety of expanded form. Overall, the expanded form is divided into two aspects: 1, study how to strengthen the Petri net modeling ability to simplify the modeling process and the model simplification, such as colored Petri nets, object-oriented Petri nets; 2, develop Petri nets analysis technology, so that more information will be get from the original system, such as the timed Petri nets, stochastic Petri nets [3-5]. In addition, people also applied IDEF0, UML method into the Petri net modeling process to further improve Petri modeling ability from different angles [6, 7].

From the current application of Petri nets, people more often used Petri net when the system is small or for a specific problem. Such as for a production scheduling

B. Hong et al. (Eds.): DASFAA Workshops 2013, LNCS 7827, pp. 220–230, 2013.

system, the model is more often describe 2-3 machine, 2-3 types of work piece and 3-5 of the specific conditions of the process. Less meet with general and can be performed on computer simulation verification of production scheduling system model [8-10], it is no common to see a general model that can be performed on computer simulation verification of production scheduling system model [8-10]. Therefore, based on the IDEF1X method to systematically design the colors set of the Petri nets, with the aid of the Danish Arhus University of the tool-CPN Tools, This paper trying to build a kind of engineering method for complex production system, lay the foundation of further system evaluation and performance analysis.

2   Color Set of the System in Colored Petri Net

Just as its name implies, Colored Petri net is the Petri net that marked the tokens in different colors, the essence of is classified tokens, so as to realize the folding of network system. A simple presentation of classification token is to use k d vector to represent a Petri net with k colors, k d vector of Each place represent the containing tokens of different colors in that place, the number of each component represents one kind of color, component value represents token number of the color.

Since Jensen used the color concept in Petri nets [11-13], puts forward the colored Petri net system, the current literature mainly concentrated in the application of colored Petri net, it is rarely to see systematic method for how to determine the color set of the colored Petri net. However, for a large scale and complicated system, it is necessary to find a systematic approach to color set in the colored Petri net and is also very helpful in the engineering application of Petri nets for system modeling.

We know, for complex system modeling, often need to build system data model and processing model. Data model describes the system state space (such as entity relationship diagram, UML class diagram, etc.), describes relationship of system activities in model Data model construct the system state space through depicting the basic entity of system and the internal relations between entities. Obviously, token and data model in a colored Petri net are associated with system entity, we can use the data model method in information system modeling to derive the color set type in colored Petri net system.

In the system modeling, when contrast Petri net model and the corresponding data model, We found place type of a Petri net is often direct an entity or several entity consisting of view in information model, a token in a place corresponding an object in entities or view, and entities properties describes the place type of necessary structure and information of Petri net model, this corresponding relations shows that we may design color set of a Petri net through the data model, corresponding relation and process see this article 4.2 and 4.3.

For the establishment of the data model, this paper applied the IDEF1X method [14]. IDEF method (IcamDEFinition Methods) is a set of system analysis and design tools planed and developed in the 1970 s by the United States air force ICAM (Integrated Computer Aided Manufacturing), it use structured approach to system analysis and design. Along with the constant derivative and extension, it has produced many different module functions at present, there are IDEF0, IDEF1, IDEF1X, IDEF2 to IDEF14, and each of them has unique functions to meet the different needs of System analysis and design. Among them, IDEF0 used to set up the system functional Modeling, IDEF1X used to set up the system data model.

IDEF1X provides a structured method to represent and support the needed information and rules of enterprise operation. One of the advantages of using IDEF1X is information can be internal sharing in enterprise. In addition, it provide the communication platform between professional and technical personnel, build consensus in the members in groups, then to establish stable data base.

3   Time Factor in Petri Net

There are 3 ways to introducing time factors in Petri nets:

1. Time correlation with transition, transition represents a certain event or operation lasting for some of time, get the Timed Transition Petri Net, TTPN. Once a transition is inspired, a certain amount of token was removed from each input place of the transition, but the transition will delay a certain period of time to put the token into the output place.
2. Time correlation with place, place represent represents a certain event or operation lasting for some of time, get the Timed Place Petri Net, TPPN. Once a transition is inspired, a certain amount of token was removed from each input place of the transition and is put into the output place, but the token can be used delay a certain period of time.
3. Time correlation with arc, that means an arc represent the process of material flow processor the process of a certain period of time. Once a transition is inspired, the token will reach the output place delay a certain period of time.

At present, most literature use the Time correlation with transition method, the reason is that Petri nets model represents a system, an event occurred (i.e. transition) in system usually needed for a certain time.

But these methods are insufficient, makes more complex relationship between the basic Petri nets and a timed Petri net. To the ways that Time correlation with transition and Time correlation with arc, in these kinds of timed Petri nets, there will be many "inaccessible" situations in accessible identification when there has no time factors in Petri nets, namely some token in the "on the way" situation. The inspired transition has removed the token from input place but not immediately put tokens in the output place so that no marking was occurred in the Petri net system and cause confusion. To the way that Time correlation with place, this kind of timed Petri nets has no effective mechanism to control the output token of time and the corresponding time delay. Therefore, based on the above reasons, this paper introduced time extension in Petri net using such a method: add a "time stamp" in token, And the time stamp value is represent according to the corresponding time delay and the time delay of output arc , the specific description as follows:

First, assumption there is a global clock to represent the time of model, time unit is according to the specific system, can be 1 second, 1 minute or a certain time. The timestamp of each token means at this time the token can be excited transition use, the running of the timed Petri net is controlled by this global clock, model Keep the current time until a new inspiring conditions occur. In timed Petri net, the inspiring condition needs not only to meet the requirements of the input token, And time stamp of token must less than or equal to the current global clock at the same time. When there is no such condition, global clock forward to the next first inspiring state. So, for conflict and complicated situation in Petri net without time factors, it will only happen in "the same time" in timed Petri net.

In timed Petri net, a time "delay" dimension is added in transition in addition to the timestamp, The "delay" dimension represent that when the transition is inspired, all the output tokens of the transition in the current model plus "delay" dimension. In this way, the additional time "delay" dimension transition does not change the nature of "instantaneous" happen, and also guarantee the output token could not be used before the timestamp, it has the same effect with defining transition inspiring time. Also, add a time "delay" dimension in the output arc, and the "delay" dimension in transition has effect on all the output tokens, but the "delay" dimension in arc just has effect on the tokens in the output arc. I transition and arc have "delay" dimension at the same time, the timestamp value of the token is the sum of the current time and the "delay" dimension in transition and arc.

There are many software support Petri net modeling and analysis [15], such as the CPN Tools established by University of Aarhus in Denmark, ProMestablished by Eindhoven University of Technology in The Netherlands, Lolaestablished by University of Rostock in German and so on. Due to the CPN Tools provides validation, simulation, hierarchical modeling and many other functions, and also support powerful Standard ML [16 -], stronger scalability and a wide range of applications, therefore, this paper use CPN Tools for Petri net modeling.

4 Production Scheduling System Modeling

4.1 Production Processing System Prototype

We know that a finished or semi-finished product is got by a series of process, and the process sequence is determined by the product technical constraints in the process designing. Generally, a process can also uses many kinds of methods to carry on the processing, but the processing results are the same. It is assumed that the working procedures that the production workshop can finish were expressed as $P_1, P_2,..., P_p$; m machines were expressed as $M_1, M_2,..., M_m$; Thus the working procedures Pi is processed by machine M_j, the unit time of the working procedure is processed as T_{ij}, it is said that a process of working procedure can be finished by many machines, a machine also can complete multiple process required processing tasks. N work pieces are expressed as $J_1, J_2,..., J_n$, work piece J_i processing process sequence (path) for J_i (P_{i1}, $P_{i2},..., P_{ix}$) , such as J_1 (P_1, P_2) represents the work piece J1 showing the motion of the first procedure for P_1, the second process for P_2;J_2 (P3,P4) said the first step of J2work piece for P_3, the second process for P4. Due to a process can be utilized in a variety of methods for processing, if we use O_{ijk} to represent the first I kind of work piece first j a process in the first k machine processing a operation, obviously, a work pieces can through the different sequence of operation to complete processing, but the processing sequence of the process is the same.

4.2   IDEF1X Model

For the production system, a workpiece is got passing through more than one processes, each process can process multiple work pieces, the corresponding relations between the workpiece and the process is many-to-many relationship. A process of processing tasks also may be completed by multiple machines, a machine can complete processing tasks required by multiple process, namely the corresponding relations between process and the machine is many-to-many relationship. The relationship between other entity in the model differ a narrative, IDEF1X model is shown as figure 1(Figure 1 in the model are only given necessary part in order to make the model simple intuitive and does not affect the discussion of the problem):

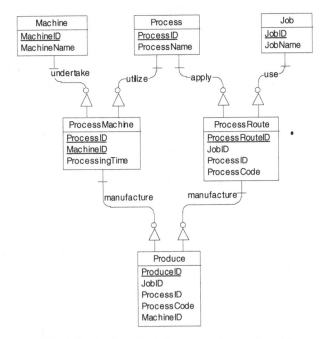

Fig. 1. Production scheduling system IDEF1X model

In figure 1, the entity "process - machine" means a process by which the machine to complete and unit processing time finishing need; Entity "routing" said production work of the specific processing process sequence; Entity "process" is a work of a process in the actual process by specific what machine finishes the actual situation.

4.3   Production Model Based on Timed Colored Petri Net

For the before production model, it is difficult to build a model by basic Petri net. Such as for a 3 machines, 2 workpiece production system (workpiece machining process and resource requirements and process time respectively see table 1 and table 2), there are 20 places, 18 transitions and 53 arcs in the basic Petri net(can see [2]). Along with the increases of machine number, number of packages and the number of work process, the basic Petri net modeling exists only in theory of feasible, and it is impossible to use this model for any simulation and analysis.

Table 1. Machining process and resource requirements

machining process	workpiece	
	J_1	J_2
1	M_1/ M_2	M_1/ M_3
2	M_2/ M_3	$M_1/ M_2/ M_3$

Table 2. Process time

Operation	$O_{1,1,1}$	$O_{1,1,2}$	$O_{1,2,2}$	$O_{1,2,3}$	$O_{2,1,1}$	$O_{2,1,3}$	$O_{2,2,1}$	$O_{2,2,2}$	$O_{2,2,3}$
Process time	3	4	3	2	4	2	3	4	4

The production model of timed colored Petri nets is shown in figure 2.According to the figure 1 IEDF1X model, entity "machine", "process", "work", "process-machine", "process route", "process" respectively derived into color set Machine, Process, Job, ProcessMachine, Task, TaskM. The Standard ML describes the colored set as follows:

- Colset Route =list STRING;
- colset Machine=INT;
- colset Process =product INT * STRING;
- colset Job =product INT * STRING;
- colset ProcessMachine=product INT*INT*INT;
- colset Task=record taID: NT*path: Route *scsl: INT*current NO:INT timed;
- colset TaskM=product Task*INT timed;

In figure 2 model, Transition "Accept" represents the workpiece design process, machining process sequence in forming the workpiece. Transition "Dispatch" represents production process. When transition "Dispatch" is inspiring, current machining is according to the token in place "Production" find the product time t in the corresponding place, a token with timestamp was produced in output place WIP with the transition time delay @+t, transition "Check" detection the completion of processing procedure: If not completed, continues to processing; if completion, then warehousing, finish all processing tasks. The meanings of place in the model in Figure 2 are shown in table 3:

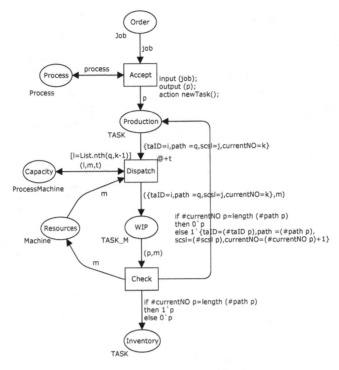

Fig. 2. Basic model of colored Petri nets

Table 3. Meanings of place in the production system

Place	Meaning	Color set
Order	Order contains workpieces	Job
Process	Enterprise standard process	Process
Production	Tasks	Task
Capacity	Produce ability	ProcessMachine
Resources	Available resources, here refers to machine	Machine
WIP	Order processing status	TaskM
Inventory	The inventory status	ProcessMachine

As the model shown in figure 2, the data in table 1 can use color set Task to represent:1`{taID=1,path=[p1,p2],scsl=0,currentNO=1}++1`{taID=2,path=[p3,p4],scsl=0,currentNO=1}; data in figure 2 can be represent by color setProcessMachine:1`(1,1,3)++1`(1,2,4)++1`(2,2,3)++1`(2,3,2)++1`(3,1,4)++1`(3,3,2)++1`(4,1,3)++1`(4,2,4)++1`(4,3,4). In Color set Task path= [p1, p2] is the color type of "list", represent the first workpiece has two processing process, the first processing procedure for process Numbers for p1 process, the second processing procedure for Numbers for p2 process. Each process can be specific "which machine to process" information from color set Process Machine, such as 1`(1,1,3) represents P1 process need Numbers 1 machine produce3 unit time,1`(1,2,4) represents P1 process need Numbers 2 machine produce4 unit time. If table 1 and table 2 contain large data (i.e., the corresponding machine number, number of packages and work process number is larger, standard data file can be storage as shown in figure 1. Petri net model in Figure 2 shows SML in simulation run-time read in the corresponding files in CPN Tools.

Figure 3 shows intuitively the corresponding relationship between the entity in IDEF1X model with the place color type in Petri net (figure with dotted line represent the corresponding relation).

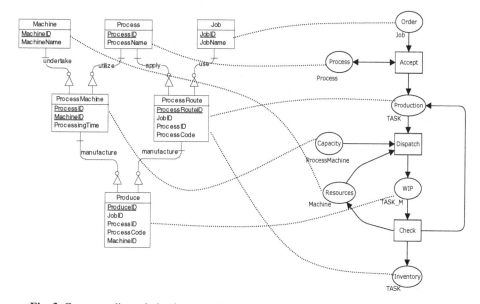

Fig. 3. Corresponding relation between data model and Petri nets in production system

Obviously, this paper established the production model has strong universality, will not change with the machine number, number of packages and the number of work process. When the machine number, packages number or work process number change, we just need to change the model of the initial value without any changes of the structure of the model. This also laid a foundation for us to deeply analysis the production organization mode of the system structure, characteristics and perfor-mance. How to use the provide functions such as Simulation in CPN Tools to verify the model, this article is no longer on detailed.

4.4   Build Production Scheduling Model

Based on the model in figure 2, using the hierarchical technology of CPN Tools, we can build the production scheduling system model as shown in figure 4.In figure 4, transition "Accept", "Plan", "Process" are substitution transition, each substitution transition has a corresponding subpage, subpage describe the substitution transition in detail.

Fig. 4. Production scheduling system model

Figure 4 shows scheduling system works as follows: First, transition "Accept" Accept the customer's order form production tasks; Then, according to the transition "Plan" and conditions on the production tasks scheduling forming production Plan; Finally, the transition "Process" processing production according to the production plan. Therefore, transition "Accept" and "Process" corresponding to the model in figure 2, and transition "Plan" is to achieve the optimal workpiece input sequence and processing given mixed the resources needed to use the function of the scheme. For the scheduling problem is very complicated [19] and is not the theme of this paper, therefore, this paper will not discuss it further. We give the very famous pattern search method A * algorithm, which is proposed to realize the function of this Petri net diagram. The subpage of transition "Plan" is shown in figure 5. Obviously, the subpage id changed when transition "Plan" using different scheduling algorithm.

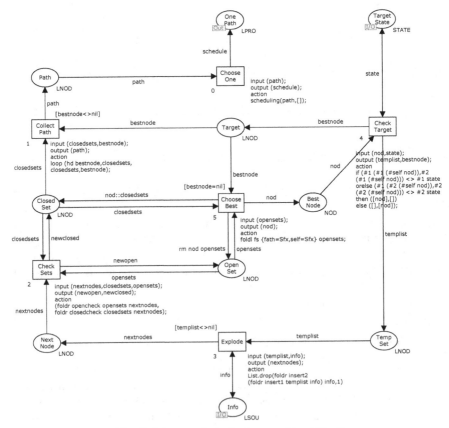

Fig. 5. Subpage of substitution transition "Plan"

5   Conclusion

The previous production scheduling systems have the positive effect on the actual application in one of the Sino-Japanese joint venture. The company is a manufacture enterprise of producing optical device. There are 3000000 products for annual output, 176 kinds of product which need 40 days for one production cycle and 94 kinds of raw materials which need 39 processes to be finished products. Because of the different of production modes, inspection standards (23 kinds of them) and production classes (6 kinds of them), there will be about 176*94*39*2*23*6=17807961 kinds of technological processes with complete combination. Obviously, Small-scale modeling way can only show problems, but not solve. This article has systematically applied IDEFIX to Petri Net modeling and built an objective system which can support the actual data for Petri Net modeling system simulation application in computer. The system laid the foundation for further simulation system performance, improving the structure of the system and efficiency.

Acknowledgements. The Subject is from Natural Science Foundation of China (71071097); and Key Discipline in Shanghai (S30504).

References

1. Wu, Z.H.: Introduction Theory to Petri Nets. Mechanical Industry Press, Beijing (2006)
2. Jiang, Z.B.: Petri Nets and Application of Model and Control in Production System. Mechanical Industry Press, Beijing (2004)
3. Su, C.: Model and Simulation in Production System. Mechanical Industry Press, Beijing (2008)
4. Jensenand, K., Kristensen, L.M.: Colored Petri Nets: Modeling and Validation of Concurrent Systems. Springer (2009)
5. van der Aalst, W., Stahl, C.: Modeling Business Processes: A Petri Net-Oriented Approach. MIT Press (2011)
6. Wei, D.G., Wu, S.L.: AUML and PETRI Nets Research of Usage in Combination. Journal of System Simulation 17(S1), 193–196 (2005)
7. Wang, Z.J., Cai, Z.X.: Based on IDEF0 Model and Indirect Petri Nets Model Methods to Study. Journal of System Simulation 8, 3915–3919 (2008)
8. Yi, S.P., Pu, J., Gao, Q.X.: Based on Petri Nets Buffer Model of Group Production Unit. Industrial Engineering and Management 12, 7–11 (2009)
9. Huang, D., Yan, J.W., Qiao, F.: Based on Petri Nets Modeling of Semiconductor Production Line. Computer Engineering 3, 69–71 (2005)
10. Jia, G.Z.: Optimization Method of Model and Simulation Petri Nets in Production System. Journal of System Simulation 8, 559–562 (2006)
11. Jensen, K.: Colored Petri Nets. Basic Concepts, Analysis Methods and Practical Use. Basic Concepts. Monographs in Theoretical Computer Science, vol. 1. Springer (1997)
12. Jensen, K.: Colored Petri Nets. Basic Concepts, Analysis Methods and Practical Use. Analysis Methods. Monographs in Theoretical Computer Science, vol. 2. Springer (1997)
13. Jensen, K.: Colored Petri Nets. Basic Concepts, Analysis Methods and Practical Use. Practical Use. Monographs in Theoretical Computer Science, vol. 3. Springer (1997)

14. Chen, Y.L.: IDEF Model Analysis and Design Method. Tsinghua University Press (2000)
15. http://www.informatik.uni-hamburg.de/TGI/PetriNets
16. Paulson, C.: ML Programming Tutorial. Mechanical Industry Press, Beijing (2005); Ke, W. (Trans.)
17. Milner, R., Tofte, M., Harper, R., MacQueen, D.: The Definition of Standard ML: Revised1997. The MIT Press (1997)
18. Ullman, J.D.: Elements of ML Programming (ML 1997edition). Prentice-Hall (1998)
19. Cai, Z.X.: Artificial Intelligence & Applications. Tsinghua University Press (2007)

Product Feature Summarization
by Incorporating Domain Information

Tao Wang, Yi Cai*, Guangyi Zhang, Yu Liu, Junting Chen, and Huaqing Min

School of Software Engineering, South China University of Technology, Guangzhou,
China
ycai@scut.edu.cn

Abstract. Feature summarization is an important problem in opinion
mining of product review. Current methods mostly cluster feature expressions with unsupervised learning methods based on lexical similarity or context information similarity. Although several semi-supervised
methods have been proposed to addressing the problem, their labeled set
needs manual definition and requires a professional knowledge to categorize features into same aspect or different aspects. In this paper, we
proposed a semi-supervised method that incorporates domain information in web sites to generate labeled set and constructs a novel context
information modeling process with EM method to solve the problem.
Experimental results show that the proposed method achieves better
performance than existing approaches.

Keywords: opinion mining, aspect extracting, machine learning.

1 Introduction

With the emergence and development of E-commerce, more and more people
purchase goods online. The amount of products and services reviews are increasing rapidly. The reviews play a important role during purchase decision making
for customer, and provide a way to know superiorities and drawbacks of products for producers. However, the huge volume of reviews makes it difficult for
users to find needed information by reading all raw reviews. Opinion mining in
product reviews is addressing a problem which includes information extraction,
summary and opinion categorization,*etc.* Feature-based opinion mining method
[2,3] is popular recently. However, there is a problem of feature overload due to
product features are identified without summarization. The extracted features
are over fined-gained (lots of them are single attributes of product's aspect, *e.g.,*
"LED", *"LCD"* are attributes of screen aspect in TV reviews) and redundancy
(i.e., many different expressions deliver same aspect, *e.g., "battery"*, *"power"* are
referring to the same aspect in camera reviews).

In this paper, we focus on the issue of feature summarabtion in feature-based
review opinion method. The domain dependence is the intractable problem in

---

* Corresponding author.

B. Hong et al. (Eds.): DASFAA Workshops 2013, LNCS 7827, pp. 231–243, 2013.

feature summary. In reviews, customers usually use different expressions to deliver the same product feature. Some work apply lexical-based methods to group the features according to lexical similarity of features in dictionary resources [1,14]. Due to some synonyms features expressions do not appear in dictionary or context-aware are considered as not lexical similar in WordNet (e.g., "interface" and "usb" in reviews about PC). Therefore, the kind of method cannot perform the task effectively. Other work relied on context information trying to address the issue [18,21]. However, most of them are unsupervised methods that may not fit use's need. Existing semi-supervised methods [20,22] need manually define labeled set. However, defining appropriate aspect category structure for a product needs a professional knowledge to identify what are important aspects and what can be categorized in same/different aspect(s) in a domain. To be feasible for massive amount of products, we generate the labeled words with aspect category information from semi-structured detail expressions in E-commerce sites.

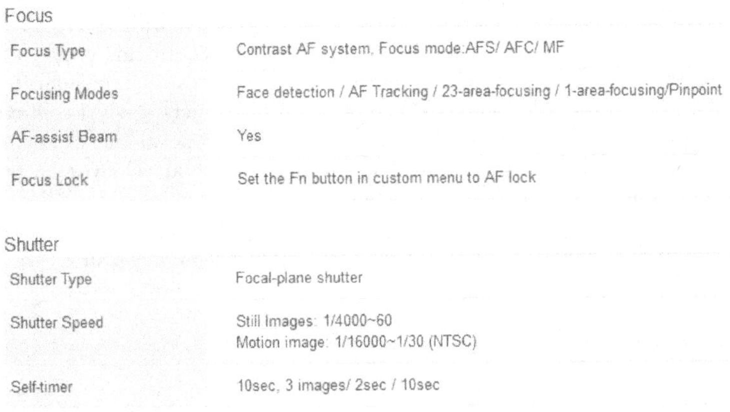

Focus	
Focus Type	Contrast AF system, Focus mode:AFS/ AFC/ MF
Focusing Modes	Face detection / AF Tracking / 23-area-focusing / 1-area-focusing/Pinpoint
AF-assist Beam	Yes
Focus Lock	Set the Fn button in custom menu to AF lock

Shutter	
Shutter Type	Focal-plane shutter
Shutter Speed	Still Images: 1/4000~60 Motion image: 1/16000~1/30 (NTSC)
Self-timer	10sec, 3 images/ 2sec / 10sec

Fig. 1. Example product detail information from Newegg.com

As an example, the detail information of a camera shown in Figure 1. The information is originally written by people, the aspect categories in it are easy to read and understand. In addition, all of those expressions are revolved around a given product. Therefore, most words used in detail expressions are the features of product. It can provide high-quality labeled set with less noise features. We select words and phrases in left side of table shown in Figure 1 as a category of aspect expressions. e.g., we can obtain two aspects from Figure 1, A_{focus} = {focus, focus type, ..., focus lock} and $A_{shutter}$ = {shutter, shutter type, ..., self-timer}.

In the context information modeling, feature distributional document that exhibit the characteristic of each feature is constructed by frequencies of context words in most current methods. Due to the huge amount of general words, such as "good", "great", exist in product reviews, some items may be clustered to same aspect for their context documents contain same general words. This kind of distributional document structure cannot describe the characteristics of features. We employ Mutual Information (MI) that measures the mutual dependence of feature and context word to construct feature distributional document and the experimental results show it can achieve a better performance.

In this paper, we employ navïe Bayesian classification in [6] to model distributional documents. Then, modify the EM algorithm to learn a classifier incorporating the domain information above. The contributions of our work are as following:

1. In the paper, we propose a semi-supervised method for product feature summarization. To get high-quality labeled set automatically, we introduce the semi-structured category information from E-commerce web sites. It does not need tedious and time consuming manual definition than current methods.
2. To describe the characteristics of features more precisely, we employ Mutual Information to construct context distributional document for each feature. In model learning, An EM algorithm based on navie Bayesian is applied to address the feature distributional document clustering, which incorporates domain information into EM learning process.
3. We conduct experiments to evaluate the proposed method with current methods, and the results show the new method outperforms the task effectively.

The remainder of the paper is organized as follows. The next section introduces previous related work in product feature clustering. The semi-supervised EM method that incorporates domain information from E-commerce web site is illustrated in Section 3. The experimental evaluation of the proposed method is introduced in Section 4. Finally, we conclude the paper in Section 5.

2   Related Work

Mining opinions from customer reviews for supporting decision analysis has been investigated by several works recently. Hu and Liu propose a system of opinion mining in [2,3].

Some existing methods are based on lexical similarity. They mostly group the features according to dictionary resources (e.g. WordNet) [1,14]. The kind of methods cannot perform very well, when some synonyms features in a specific domain do not appear in dictionary or they are not be lexical similar in WordNet. Additionally, some synonyms words in WordNet may not synonyms in given domain, *e.g.* "opinion" and "view" in camera product review corpus.

Another kind of methods is distributional similarity which is popular in text classification. Those methods are assumed that context words of features belonged to same aspect are similar. Some similarity measures, such as *Euclidean distance, Cosine, Jaccard, etc* [15], are used to calculate the similarity of documents. Some other work also computes the Pointwise Mutual Information (PMI)

[19] between words and then clusters words with K-Means based on cosine similarity of PMI weight. The work in [5] uses two constraints based on sharing words and lexical similarity to generate labeled data.

Some works also employ topic model [16] for the problem recently. [17] states a multilevel latent semantic association (LaSA) model to categorize product feature. The first LaSA model is to group words into a set of concepts according to their virtual context documents. The result generates the latent semantic structure for each product feature. Then, the product features are categorized according to their latent semantic structures. [20] presents a boot-strapping algorithm for aspect related words extraction based on labeled set. Arjun proposes a statistical model to extract and categorize aspect according to some labeled words for each aspect category in [22]. The labeled words in both of them are manually defined.

3   The Proposed Method

The proposed method is semi-supervised, however, it does not need manual input by introducing details expressions in E-commerce web site like *newegg.com* and *ebay.com* and so on. Those detail information which includes feature expressions and aspect categories is applied to generate labeled features in our semi-supervised method. Afterwards, we extract frequent nouns or noun phrases as candidate features. We extract contexts for both labeled features and unlabeled features from review corpus and calculating Mutual Information (MI) to get the distributional document of each feature. After those procedures, we employ EM learning method based on reformative navïe Bayesian to cluster features according their distributional documents.

3.1   Labeled Set Generation

Generally, semi-supervised method needs some labeled set as guide of its learning process, and it is significant to choose appropriate labeled set. The detail expressions in E-commerce web site are originally wrote by people, the category structure implied in them is easy to read and understand. In addition, all of those expressions are revolved around a given product. Therefore, the words used in detail expressions are mostly ground truth features of product, which can provide high-quality labeled set with less noise features.

Although detail information is a good labeled set source, the detail expressions are still not good enough to be labeled set because most those expressions are noun phrases. They are too professional expressions for general consumers. This kind of professional expression appears with low frequency in reviews due to most customers prefer to use some more daily expressions. There are little contexts about those detail expressions in our observation, and it will influence performance of semi-supervised learning. For example, *"LED"* used to describe the screen of TV will be assigned in unlabeled set because there is only *"LED technology"* in detail expressions. In addition, there are some general words, such as *"type"*, *"mode"*, appearing in several different aspects. The quality of labeled set will decrease when general words without aspect orientations. To filter general words and not destroy the meaning of detail expressions, we propose an algorithm to extract keywords of detail phrases. Keywords can gain sufficient context information about detail expressions. The keywords of expression phrases are words that frequently appear in current aspect detail expressions, rarely appear in other aspect detail expressions. The algorithm is given in **Algorithm 1**.

Input : A collection of detail expression aspects $A : \{A_1, ..., A_K\}$, a set of expression words $W : \{w_1, ..., w_V\}$
Output: Keywords classes of detail expression $S : \{s_1, ..., s_K\}$

```
 1  for each word w_t ∈ W do
 2  │    F_ti ← frequency in each aspect A_i;
 3  │    F_t  ← frequency in all aspects A;
 4  end
 5  for each aspect A_i ∈ A do
 6  │    for each expression P_j ∈ A_i do
 7  │    │    s_i.add(P_j);
 8  │    │    for each word w_t ∈ P_j do
 9  │    │    │    if w_t is unique word with max F_ti in P_j then
10  │    │    │    │    s_i.add(w_t);
11  │    │    │    else
12  │    │    │    │    if w_t is unique word with min (F_t − F_ti) in P_j then
    │    │    │    │         s_i.add(w_t);
13  │    │    │    end
14  │    │    │    if (F_t == F_ti) then  s_i.add(w_t);
15  │    │    end
16  │    end
17  end
```

Algorithm 1. Aspect Keywords Extraction

In Algorithm 1, $\{A_1, ..., A_K\}$ are the aspects derived from E-commerce sites, $\{P_1, ..., P_n\}$ are detail expression items in each A_i, all of those expressions are delivered by word w_t in vocabulary $\{w_1, w_2, ..., w_V\}$. C is keyword collection. Firstly, counting the local frequency F_{ti} of word w_t within each aspect A_i and global frequency F_t of each word w_t in all aspects (lines 1 to 4). In second step (line 5 to 17), each detail expression is represented by keywords. Then, we select the word in expression phrase, and we find unique words with max frequency in current aspect (lines 9,10); otherwise, adding the word in expression phrase which is the unique word with min frequency in other aspects (line 12), that is, min $(F_t − F_{ti})$. For the situation that each word's local frequency is equal to global frequency in P_j, if the words only appear in one aspect, they will be added into keyword set (line 14); otherwise, it has to strictly match whole words in expression to identify its aspect orientation. The output C will be used as labeled data in semi-supervised method.

3.2 Feature Candidates Extraction

The method of this step includes finding frequent nouns and noun phrases, extracting opinion words around frequent terms, then appending nouns and noun phrases modified by opinion words in feature set. There will be a large number of feature candidates and some of them may be noise.

A frequent noun or noun phrase may be a ground-truth feature about a given product, it also may be an idiom usually used by people. This brings some frequent noise in the candidates set. However, ground-truth features are mostly modified by different opinion words because different customers have different demands and tastes about same product or same feature. In contrast, an idiom is used in a fixed and traditional way. To eliminate frequent idioms, we filter the nouns or noun phrases with number of modified opinion words less than a threshold. In the paper, we set the threshold as 3. Due to detail information can provide high-quality labeled seed words and understandable category, the proposed method can identify proper category.

3.3 Obtaining Feature Distributional Document

To overcome the domain dependence of feature in reviews, we assume that words with similar meaning tend to appear in similar contexts [4]. Therefore, we firstly extract the contexts about each feature in labeled set and unlabeled set. In our experiments, we use 7 as the size of the context windows. However, only using frequencies of context words to weight their correlations toward feature expressions during model learning in [5,6] is not enough. For the reason that these some general words may be in different feature contexts, such as, *"good"* and *"great"*. Unfortunately, this kind of words is very widespread in context expressions. It can cause that some features to be clustered to a same aspect for their context documents contain some same general words. It demands an appropriate correlation weight method that can exhibit the characteristic of each feature expression in distributional document.

We adopt a statistic correlation calculation method, i.e., mutual information (MI), to find relevant and discriminative context expressions for each feature expression:

$$\mu(w_t, d_i) = MI_{normal}(w_t, f_i) = \frac{MI(w_t, f_i)}{\sum_{s=1}^{|d_i|} MI(w_s, f_i)} \tag{1}$$

$$MI(w_t, f_i) = Pr(w_t, f_i) log_2\left(\frac{Pr(w_t, f_i)}{Pr(w_t)Pr(f_i)}\right) \tag{2}$$

where $\mu(w_t, d_i) \in [0, 1]$ is the correlation weight variable to estimate the weight of word w_t in context document of feature expression f_i. $|d_i|$ is the number of context words surround feature expression f_i. $Pr(w_t, f_i)$ is the joint probability of both terms appearing in one text window, and $Pr(w)$,$Pr(f)$ are the marginal probability of context word and feature expressions respectively. So we can get feature distributional document $d_i = \{w_1 : \mu(w_1, d_i), ..., w_{|d_i|} : \mu(w_{|d_i|}, d_i)\}$.

3.4   Semi-supervised Learning

A semi-supervised learning method based reformative navïe Bayesian is proposed to assign aspect category label to unlabeled data with the feature distributional documents. The distributional document of each feature expression data are constructed by context words surround the feature expressions and their correlation weights discussed in the previous section. That is, the feature expression f_i is replaced by its distributional document d_i in the learning step. According the relationship of documents, the semi-supervised method tries to discover domain dependent synonyms in product reviews.

We choose the navïe Bayesian to construct the document model for text classification. Navïe Bayesian assumes that word independence so the model to be characterized with a greatly reduced number of parameters [10,11]. Given a document d_i which contains a word sequence $\{w_{d_i,1}, ..., w_{d_i,|d_i|}\}$, the $w_{d_i,k}$ means the kth word w in document d_i. After introducing the aspect category information, the document is considered to be generated by several mixture aspect components (c_j). The conditional probability of a document given a mixture aspect component (θ is the model parameter set):

$$P(d_i|c_j; \theta) = P(w_{d_i,i}, ..., w_{d_i,|d_i|}|c_j; \theta) \tag{3}$$

Then, we make an navïe Bayesian assumption: the words in document are independent with others, the positions of words are independent for the document distribution. So *Equation 3* can be relaxed to:

$$P(d_i|c_j; \theta) = \prod_{k=1}^{|d_i|} P(w_{d_i,k}|c_j; \theta) \tag{4}$$

The goal of the proposed model is to assign aspect classes for documents, so the conditional probability of a component in a given document can be deduced by Bayesian theory as follows:

$$P(c_j|d_i; \theta) = \frac{P(c_j) \prod_{k=1}^{|d_i|} P(w_{d_i,k}|c_j; \theta)}{\sum_{r=1}^{|C|} P(c_r) \prod_{k=1}^{|d_i|} P(w_{d_i,k}|c_r; \theta)} \tag{5}$$

Traditionally, the navïe Bayesian is used to classify when a classifier is learned after training with labeled data. As described above, the labeled data from detail expression are ground-truth, but those labeled data are not sufficient to learn a proper parameters.

To solve the problem, we apply EM algorithm to navïe Bayesian for calculating expectations of aspect class labels of unlabeled documents, $P(c_j|d_i; \theta)$. EM will perform unsupervised clustering when labeled and unlabeled documents are imported in navïe Bayesian model simultaneously. Although it can be initialized using labeled document information, the original intention is to assign the unlabeled documents while the labeled documents act as guides during model learning. We modify the EM algorithm in [6] by making a distinction between labeled set and unlabeled set during the EM learning. To calculate the $P(c_j|d_i; \theta)$ iteratively, it needs to calculate two probabilities.

The estimated word probability $P(w_t|c_j)$ is:

$$P(w_t|c_j; \hat{\theta}) = \frac{\frac{1}{|V|} + \sum_{i=1}^{|D|} \Lambda(i)\mu(w_t, d_i)P(c_j|d_i; \hat{\theta})}{1 + \sum_{s=1}^{|V|} \sum_{i=1}^{|D|} \Lambda(i)\mu(w_s, d_i)P(c_j|d_i; \hat{\theta})} \tag{6}$$

where $\hat{\theta}$ is the estimated model parameters and illustrates the state change of model parameters during iterative update compared with θ in *Equation 5*. Due to the number of labeled data is small in comparison to the unlabeled documents. Large number of unlabeled documents may product aspect classes which are not associated to labeled classes. $\Lambda(i)$ is credible weight parameter aiming to introduce supervised into unsupervised EM learning.

$$\Lambda(i) = \begin{cases} 1, & \text{if } d_i \in D^l \\ \lambda, & \text{if } d_i \in D^u \end{cases} \tag{7}$$

If document d_i is in labeled document collection. $\Lambda(i)$ will be 1. Otherwise, $\Lambda(i)$ will be λ in the range of $\in (0,1)$ when d_i is a member of unlabeled document collection. $\mu(w_t, d_i)$ is correlation weight of word w_t in vocabulary $V = \{w_1, ..., w_{|V|}\}$ for document d_i in document set $D = \{d_1, ...d_{|D|}\}$. It is valued by the frequency of word w_t occuring in document d_i in [5]. In the proposed method, $\mu(w_t, d_i)$ is valued by normalized Mutual Information between w_t and f_i.

The aspect prior probabilities $P(c_j)$ is:

$$P(c_j|\hat{\theta}) = \frac{\frac{1}{|C|} + \sum_{i=1}^{|D|} \Lambda(i)P(c_j|d_i; \hat{\theta})}{1 + \sum_{r=1}^{|C|} \sum_{i=1}^{|D|} \Lambda(i)P(c_r|d_i; \hat{\theta})} \tag{8}$$

where $C = \{c_1, ..., c_{|C|}\}$ is the set of aspect category and $|C| = K$, K is umber of product detail expressions.

It should be noted that Laplace smoothing is widely used in navïe Bayesian [5] *et.al*. Due to $\mu(w_s, d_i)$ is normalized Mutual Information $\in [0, 1]$ in proposed method. Increasing 1 in Laplace smoothing is large comparing with maximum likelihood probability. In this method, add $\frac{1}{|V|}$ and $\frac{1}{|C|}$ smoothing is used respectively in *Equations 6 and 8*. The EM algorithm of our method is given in **Algorithm 2**.

4   Experimental Results

In this section, we conduct experiments to compare our method with other existing methods for solving the product feature summarization problem.

Input   : Labeled distributional documents D_l and Unlabeled distributional documents D_u

Output: A classifier, $\hat{\theta}$, that assign aspect labels for unlabeled documents

1 **Initialize:** *Set credible weight λ of D_u, Build an initial naive Bayesian classifier from the D_l using Equation 6 and 8.*;

2 **repeat**

3   // E-Step

4   **for** *each document $d_i \in D_l \bigcup D_u$* **do**

5   *Use the current classifier, $\hat{\theta}$, to compute each document $P(c_j|d_i)$ by Equation 5.*

6   **end**

7   // M-Step

8   *Re-estimate the classifier, $\hat{\theta}$, by computing $P(w_t|c_j)$ and $P(c_j)$ with Equation 6 and 8.*

9 **until** *the classifier parameters convergence*;

Algorithm 2. EM Algorithm for Navïe Bayesian Model

4.1 Data Set and Preprocessing

We crawls reviews of 5 popular products and their detail expressions, they are from six domain: TV, game player, phone, tablet PC and camera. All the data set is from *www.amazon.com* and *www.newegg.com*. We first perform simple pre-processing on these reviews: (1) transform words into low case; (2) use the NL-Processor linguistic parser (NLProcessor 2000) to yield part-of-speech tag same as in [2]; (3) remove punctuations, stop words; (4) stemming each word with MIT Java Wordnet Interface (JWI)[1]. To obtain gold standard, we hand-annotated the categories of the candidate features before semi-supervised learning. The details of data set are illustrated in Table 1.

Table 1. Evaluation Corpus Statistics

	TV	Gamer	Phone	T-PC	Camera
Sentences	4984	20398	9099	5617	4011
Reviews	300	2287	737	326	482
Features	154	240	238	162	173
Groups	8	7	10	12	13

4.2 Evaluation

To evaluate the proposed method, we employ two common measures for evaluating clustering in [5,13], *Entropy* and *Purity*. We first briefly describe the

---

[1] http://projects.csail.mit.edu/jwi/

definitions of *entropy* and *purity*. Given a data set D, its gold partition is $G = \{g_1, ..., g_k\}$, where k is the given number of clusters. The clustering method also produces k clusters, that partition D into k disjoint subsets, $\{D_1, ..., D_k\}$.

Entropy: The smaller value of entropy is better. For each cluster, its entropy can be measured as *Equation 9*, where $Pr_i(g_j)$ is the proportion of gold partition g_j in generated cluster subset D_i. The total entropy of the whole clustering is calculated as *Equation 10*:

$$entropy(D_i) = -\sum_{j=1}^{k} Pr_i(g_j)log_2 Pr_i(g_j) \tag{9}$$

$$entropy_{total}(D) = \sum_{j=1}^{k} \frac{|D_i|}{|D|} entropy(D_i) \tag{10}$$

Purity: This measures the extent that a cluster contains only one gold partition. The cluster purity is calculated by *Equation 11* and the total purity of the whole clusters is computed with *Equation 12*

$$purity(D_i) = max_j(Pr_i(g_j)) \tag{11}$$

$$purity_{total}(D) = \sum_{j=1}^{k} \frac{|D_i|}{|D|} purity(D_i) \tag{12}$$

4.3   Baseline

The proposed algorithm is compared with the L-EM proposed by *Zhai et.al.* in 2011 [5], and they show that their L-EM algorithm outperforms existing methods include: K-means clustering series, topic modeling series, correlation series, lexical similarity series, and EM classification series. That method uses two soft constraints to generate labeled set then apply EM to learning navïe Bayes classifier. Here, we choose it as our compared baseline method. The feature expressions used are the same in our method and L-EM.

4.4   Results

Comparative Performance: *Fig.2* presents the evaluation result of our method described as MI-EM and L-EM. As it shown, our model outperforms L-EM in there aspects: Entropy, Purity and Runtime. Top two sub-figures in Fig.2 show that the entropy and purity of MI-EM (our method) compared to L-EM (zhai's), MI-EM outperforms the L-EM due to the noise in extracted feature set. The L-EM needs to a precise feature set since labeled set in the method is directly generated from extracted feature set. The discrepancy of runtime shown in bottom figure between two method is very sharp, because the L-EM method needs to calculate the lexical similarity of words or noun phrase from WordNet to generate labeled set while no need in MI-EM.

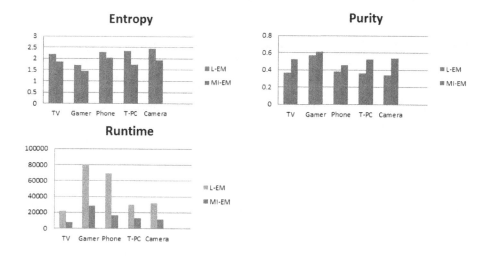

Fig. 2. Evaluation result of our method described as MI-EM and L-EM

Impact of MI: *Fig.3* shows that the performances of normal EM method that uses TF as the correlation weight of word in distributional document compared with MI-EM. It states that incorporating Mutual Information can improve the performance of classification method. The two sub-figures in Fig.3 shows that the entropy and purity of EM (TF mode) and MI-EM (MI mode) in the condition of different unlabeled credible weight λ. It can be concluded: **1.** Set λ to value between 0 and 1 can outperforms than eight $\lambda = 0$ or $\lambda = 1$, it indicates that the cluster in unlabeled set can result in poor classification. **2.** The best entropy and purity of MI-EM are better than EM. It illustrates that feature distributional documents constructed by MI can improve the classification due to MI can characterize feature expressions more precisely. As described in [6], the λ which achieves the best classification is larger when smaller labeled data set. It will be smaller when there is enough labeled data. In practice the value of the tuning parameter λ can be selected by cross-validation.

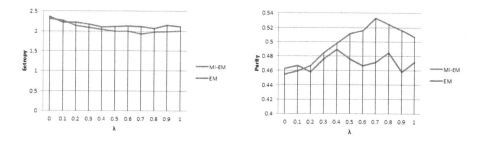

Fig. 3. Performances of MI-EM

5   Conclusion

This paper introduce a novel method to cluster product feature for review mining applications. we propose a semi-supervised method to resolve the problem. The approach incorporates domain information which is widespread in web site, and can generate labeled set for semi-supervised learning. Then we proposed amendatory EM method with Mutual Information to improve performance. The experiments show our method is better than compared baseline methods in evaluation measures.

Acknowledgements. This work is supported by Foundation for Distinguished Young Talents in Higher Education of Guangdong, China (NO. LYM11019); the Guangdong Natural Science Foundation, China (NO. S2011040002222); the Fundamental Research Funds for the Central Universities, SCUT (NO. 2012ZM0077); National University Innovation Research and Training Program (101056137, 20111056172, 201210561104, 201210561106, 201210561108); and Guangdong Province University Innovation Research and Training Program (1056112107, 1056112109, 1056112111).

References

1. Liu, B., Hu, M., Cheng, J.: Opinion Observer: Analyzing and Comparing Opinions on the Web. In: WWW (2005)
2. Hu, M., Liu, B.: Mining Opinion Features in Customer Reviews. American Association for Artificial Intelligence (2004)
3. Hu, M., Liu, B.: Mining and Summarizing Customer Reviews. In: KDD (2004)
4. Harris, Z.S.: Mathematical structures of language. Interscience Tracts in Pure and Applied Mathematics (1968)
5. Zhai, Z., Liu, B., Xu, H., Jia, P.: Clstering Product Features for Opinion Mining. In: WSDM (2011)
6. Nigam, K., Mccallum, A.K., Thrun, S., Mitchell, T.: Text Classification from Labeled and Unlabeled Documents using EM. Machine Learning (2009)
7. Peter, F., Brown, V.J., Della Pietra, V.J., Della Pietra, J.C., Lai, R.L.: Mercer: Class-based n-gram models of natural language. Association for Computational Linguistics (1992)
8. Lin, D., Wu, X.: Phrase Clustering for Discriminative Learning. In: ACL (2009)
9. Matsuo, Y., Sakaki, T., Uchiyama, K., Ishizuka, M.: Graph-based Word Clustering using a Web Search Engine. EMNLP (2006)
10. Lewis, D.D.: Naive (Bayes) at forty: The independence assumption in information retrieval. In: Nédellec, C., Rouveirol, C. (eds.) ECML 1998. LNCS, vol. 1398, pp. 4–15. Springer, Heidelberg (1998)
11. McCallum, A., Nigam, K.: A comparison of event models for naive Bayes text classification. AAAI (1998)
12. McCallum, A.K., Nigam, K.: Employing EM and Pool-Based Active Learning for Text Classification. In: ICML (1998)
13. Liu, B.: Web data mining: Exploring hyperlinks, contents, and usage data. Springer (2006)

14. Carenini, G., Ng, R.T., Zwart, E.: Extracting Knowledge from Evaluative Text. In: K-CAP (2005)
15. Lee, L.: Measures of Distributional Similarity ACL (1999)
16. Blei, D., Ng, A., Jordan, M.: Latent dirichlet allocation. Journal of Machine Learning Research (2003)
17. Guo, H., Zhu, H., Guo, Z., Zhang, X., Su, Z.: Product Feature Categorization with Multilevel Latent Semantic Association. In: CIKM (2009)
18. Patrick, P., Eric, C., Arkady, B., Ana-Maria, P., Vishnu, V.: Web-scale distributional similarity and entity set expansion. In: ACL (2009)
19. Popescu, A.-M., Etzioni, O.: Extracting Product Features and Opinions from Reviews. In: HLT-EMNLP (2005)
20. Wang, H., Lu, Y., Zhai, C.: Latent aspect rating analysis on review text data: a rating regression approach. In: SIGKDD (2010)
21. Thad, H., Daniel, R.: Lexical semantic relatedness with random graph walks. EMNLP (2007)
22. Mukherjee, A., Liu, B.: Aspect extraction through Semi-Supervised modeling. In: ACL (2012)

Author Index